PAINLESS
Algebra

Lynette Long, Ph.D.

illustrated by Hank Morehouse

BARRON'S

All inquiries should be addressed to:
Barron's Educational Series, Inc.
250 Wireless Boulevard
Hauppauge, New York 11788
http://www.barronseduc.com

Library of Congress Catalog Card No.: 98-18911
International Standard Book No. 0-7641-0676-7

Long, Lynette.
    Painless algebra / Lynette Long ; illustrated by Hank Morehouse.
        p.   cm.
    Includes index.
    Summary: Introduces the language, number systems, integers, and different types of problems of algebra, including variables, quadratic equations, exponents, roots, and radicals.
    ISBN 0-7641-0676-7
    1. Algebra. [1. Algebra]   I. Morehouse, Hank, ill.   II. Title.
QA152.2.L66   1998
512—dc21   98-18911
                                                                CIP
                                                                AC

PRINTED IN CANADA
30 29 28 27 26 25 24 23 22 21 20

CONTENTS

INTRODUCTION

Painless Algebra! Impossible, you think. Not really. I have been teaching math or teaching teachers how to teach math for over twenty years. Math is easy. You just have to remember that math is a foreign language like French, German, or Spanish. Once you understand "Math Talk" and can translate it to "Plain English," Algebra really is Painless.

Painless Algebra has some unique features that will help you along the road to success. First, it has "Math Talk!" boxes, which will teach you how to change "Math Talk" to "Plain English." You will also find boxes called "Caution—Major Mistake Territory!" which will help you avoid common pitfalls. There are also "Reminder" boxes to help you remember what you may have forgotten. And of course, there are those dreaded "Word Problems," but I've solved them all for you, so it's painless.

Chapter One is titled "A Painless Beginning," and it really is. It is an introduction to numbers and number systems. It will teach you how to perform simple operations both on numbers and variables painlessly, and by the end of the chapter you will know what "Please Excuse My Dear Aunt Sally" means.

Chapter Two shows you how to add, subtract, multiply, and divide both positive and negative numbers. It's painless. The only trick is remembering which sign the answer has, and with a little practice you'll be a whiz.

Chapter Three teaches you how to solve equations. Think of an equation as a number sentence that contains a mystery number. All you have to do is figure out the value of the mystery number. It's painless. Just follow a few simple steps to success.

Chapter Four shows you how to solve inequalities. What happens when the mystery number is not part of an equation but instead is part of a number sentence in which one part of the sentence is greater than the other part of the number sentence? Now what could the mystery number be? Does it sound complicated? Trust me. It's painless.

Chapter Five is on exponents. What happens when you multiply a number by itself seven times? You can write 2 times 2 times 2 times 2 times 2 times 2 times 2, or you can use an exponent and write 2^7, which is two to the seventh power. This chapter will introduce you to exponents, which are amazing shortcuts, and teach you how to work with them.

Chapter Six is about roots and radicals. Roots are not tree roots, but the opposite of exponents. What are radicals? You'll have to wait until Chapter Six to find out.

Chapter Seven shows you how to solve quadratic equations. "Quadratic" is a big word, but don't get nervous. It's just a fancy name mathematicians give to an equation with an x-squared term in it. It's really painless.

Chapter Eight shows you how to solve systems of equations. Systems of equations are two or more equations taken together. You try to find a single answer that will make them all true. You'll learn to solve systems in many different ways. It's fun, since no matter how you solve them you always get the same answer. But that's one of the magic things about mathematics!

If you are learning algebra for the first time, or if you are trying to remember what you learned but forgot, this book is for you. It is a painless introduction to algebra that is both fun and instructive. Plunge forward to the first page. There's nothing to be afraid of. Remember, it's painless.

CHAPTER ONE

A Painless Beginning

Algebra is a language. In many ways it is just like learning French, Italian, or German, or maybe even Japanese. In order to understand Algebra, you need to learn how to read Algebra and how to change Plain English to Math Talk and Math Talk back to Plain English.

In algebra, a letter is often used to stand for a number. The letter used to stand for a number is called a *variable*. You can use any letter, but a, b, c, n, x, y, and z are the most commonly used letters. In the following sentences, different letters are used to stand for numbers.

$$x + 3 = 5 \qquad x \text{ is a variable.}$$
$$a - 2 = 6 \qquad a \text{ is a variable.}$$
$$y \div 3 = 4 \qquad y \text{ is a variable.}$$
$$5z = 10 \qquad z \text{ is a variable.}$$

When you use a letter to stand for a number, you don't know what number the letter is standing for. Think of the letters x, a, y, or z as mystery numbers.

Variables can be part of an expression, an equation, or an inequality. A mathematical expression is part of a mathematical sentence, just like a phrase is part of an English sentence. Here are a few examples of mathematical expressions: $3x$, $x + 5$, $x - 2$, $x \div 10$. In each of these expressions it is impossible to know what x is. The variable x could be any number.

Mathematical expressions are named based on how many terms they have. *Monomial* expressions have one term.

x is a monomial expression.
3 is a monomial expression.
z is a monomial expression.
$6x$ is a monomial expression.

Binomial expressions have two unlike terms combined by an addition or subtraction sign.

$x + 3$ is a binomial expression.
$a - 4$ is a binomial expression.
$x + y$ is a binomial expression.

Trinomial expressions have three unlike terms combined by an addition or subtraction sign.

$x + y - 3$ is a trinomial expression.
$2x - 3y + 7$ is a trinomial expression.
$4a - 5b + 6c$ is a trinomial expression.

Polynomial expressions can have two, three, or more unlike terms combined by an addition or subtraction sign. Binomials, and trinomials are all polynomial expressions. The following are also polynomial expressions.

$x + y + z - 4$ is a polynomial expression.
$2a + 3b - 4c + 2$ is a polynomial expression.

A *mathematical sentence* contains two mathematical phrases joined together by an equals sign or an inequalities sign. An *equation* is a mathematical sentence in which the two phrases are joined together by an equals sign. Notice how the word *equation* starts the same way as the word *equal*.

$3 + 6 = 9$ is an equation.
$x + 1 = 2$ is an equation.
$7x = 10$ is an equation.
$0 = 0$ is an equation.
$4x + 3$ is not an equation. It does not have an equals sign. It is a mathematical expression.

Some equations are true and some equations are false.

$3 + 5 = 7$ is an equation, but it is false.
$1 = 5$ is also an equation, but it is also false.
$3 = 2 + 1$ is a true equation.
$x + 1 = 5$ is an equation. It could be true and it could be false.

Whether $x + 1 = 5$ is true or false depends on the value of x. If x is 4, the equation $x + 1 = 5$ is true. If $x = 0$, $x + 1 = 5$ is false. If x is any number besides 4, the equation $x + 1 = 5$ is false.

An *inequality* is a mathematical sentence in which two phrases are joined by an inequality symbol. The inequality symbols are *greater than*, ">," *greater than or equal to*, "≥," *less than*, "<," and *less than or equal to*, "≤."

Six is greater than five is written as $6 > 5$. Seven is less than ten is written as $7 < 10$.

MATHEMATICAL OPERATIONS

In mathematics, there are four basic operations: addition, subtraction, multiplication, and division. When you first learned to add, subtract, multiply, and divide, you used the symbols $+$, $-$, $\times$, and $\div$. In algebra, addition is still indicated by the plus ($+$) sign and subtraction is still indicated by the minus ($-$) sign.

MATH TALK!

Watch how these mathematical expressions are changed from Math Talk into Plain English

$$5 + 1$$
five plus one

$$a + 6$$
a plus six

MATH TALK!

Watch how these mathematical expressions are changed from Math Talk into Plain English

$$4 - 3$$
four minus three

$$a - 5$$
a subtract five

Addition

When you add you can add only *like terms*.

All numbers are like terms.
5, 3, 0.4, and $\frac{1}{2}$ are like terms.

Variables that use the same letter are like terms.
$3z$, $-6z$, and $\frac{1}{2}z$ are like terms.

A number and a variable are *unlike terms*.

7 and x are unlike terms.

Variables that use different letters are unlike terms.

$3z$, b, and $-2x$ are unlike terms.

You can add any numbers together.

$$3 + 6 = 9$$
$$5 + 2 + 7 + 6 = 20$$

You can also add variables together as long as they are the same variable. You can add x's to x's and y's to y's, but you cannot add x's and y's together. To add like variables, just add the *coefficients*. The coefficient is the number in front of the variable.

In the expression $7a$, 7 is the coefficient and a is the variable.
In the expression $5x$, 5 is the coefficient and x is the variable.
In the expression $\frac{1}{2}y$, $\frac{1}{2}$ is the coefficient and y is the variable.
In the expression x, 1 is the coefficient and x is the variable.

Now watch how these like terms are added together.

$$3x + 7x = 10x$$
$$4x + 12x + \frac{1}{2}x = 16\frac{1}{2}x = \frac{33}{2}x$$
$$4a + a = 5a$$

You cannot simplify $3x + 5y$ because the variables are not the same.

You cannot simplify $3x + 4$ because $3x$ is a variable expression and 4 is a number.

Caution—Major Mistake Territory!

When adding expressions with variables, just add the coefficients and attach the variable to the new coefficient. Do not put two variables at the end.

$$2x + 3x \neq 5xx$$
$$2x + 3x = 5x$$

Note: An equal sign with a slash through it ($\neq$) means not equal to.

Subtraction

You can also subtract like terms.

You can subtract one number from another number.

$$7 - 3 = 4$$
$$12 - 12 = 0$$

You can subtract one variable expression from another variable expression as long as they contain the same variable. Just subtract the coefficients and keep the variable the same.

$$7a - 4a = 3a$$
$$9x - 2x = 7x$$
$$3x - x = 2x \text{ (Remember, the coefficient of } x \text{ is 1.)}$$
$$4y - 4y = 0y = 0$$

You cannot simplify $3x - 4y$ because the terms do not have the same variable. You cannot simplify $100 - 7b$ because 100 and $7b$ are not like terms. You cannot simplify $5x - 3y$ because $5x$ and $3y$ are not like terms.

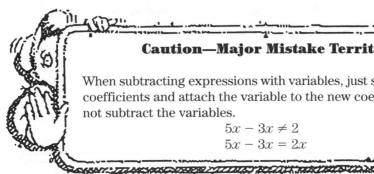

Caution—Major Mistake Territory!

When subtracting expressions with variables, just subtract the coefficients and attach the variable to the new coefficient. Do not subtract the variables.

$$5x - 3x \neq 2$$
$$5x - 3x = 2x$$

BRAIN TICKLERS
Set # 1

Add and subtract these algebraic expressions.

1. $3x + 7x$ 5. $6x - 4x$

2. $4x + x$ 6. $3x + 2$

3. $3x - 3x$ 7. $10 - 4x$

4. $10x - x$

(Answers are on page 32.)

Multiplication

An $\times$ is seldom used to indicate multiplication. It is too easy to confuse "$\times$," which means multiply, with "x" the variable. To avoid this problem mathematicians use other ways to indicate multiplication. Here are three ways to write "multiply."

1. A dot $\cdot$ is used to write "multiply."
$$3 \cdot 5 = 15$$

2. Writing two letters or a letter and a number next to each other is another way of saying "multiply."
$$7b = 7 \cdot b$$

3. Writing a letter or a number next to a set of parentheses says "multiply."
$$6(2) = 12$$

MATH TALK!

Watch how these mathematical expressions are changed from Math Talk into Plain English

$$3 \times 4$$
three times four

$$5 \cdot 2$$
five times two

$$6(2)$$
six times two

$$4x$$
four x

$$5y$$
five times y

You can multiply like and unlike terms together.

You can multiply any two numbers together.
$$3(4) = 12$$
$$8\left(\frac{1}{2}\right) = 4$$

You can multiply any two variables together.
$$x \cdot x = (x)(x) = x^2$$
$$y \cdot y = (y)(y) = y^2$$
$$a \cdot b = (a)(b) = ab$$
$$x \cdot y = (x)(y) = xy$$

You can multiply a number times a variable.
$$3 \cdot x = 3x$$
$$7 \cdot y = 7y$$

You can even multiply two expressions together if one is a number and the other is a variable with a coefficient. To multiply these expressions together it takes two *painless* steps.

1. Multiply the coefficients together.

2. Put the variable on the end of the answer.

Here are some examples:

3 times 5x
First multiply the coefficients together.
$$3 \cdot 5 = 15$$
Next put the variable on the end of the answer.
$$15x$$
Here is the problem and the answer.
$$3 \cdot 5x = 15x$$

6y times 2
First multiply the coefficients together.
$$6 \cdot 2 = 12$$
Next put the variable on the end of the answer.
$$12y$$
Here is the problem and the answer.
$$6y \cdot 2 = 12y$$

You can also multiply two expressions together that each have both numbers and variables. In order to multiply these expressions together it takes three steps.

1. Multiply the coefficients together.

2. Multiply the variables together.

3. Combine the two answers.

Here are a couple of examples.

Multiply 3x times 2y.
First multiply the coefficients together.
$$3 \cdot 2 = 6$$
Next multiply the variables together.
$$x \cdot y = xy$$
Combine the two answers by multiplying them together.
$$6xy$$
Here is the problem and the answer.
$$3x \cdot 2y = 6xy$$

Multiply 4x times 5x.
First multiply the coefficients.
$$4 \cdot 5 = 20$$
Next multiply the variables.
$$x(x) = x^2$$
Combine both answers by multiplying them together.
$$20x^2$$
Here is the problem and the answer.
$$4x \cdot 5x = 20x^2$$

Multiply 6x times y.

First multiply the coefficients. The coefficients are 6 and 1.

$$6 \cdot 1 = 6$$

Next multiply the variables.

$$x \text{ times } y = xy$$

Combine both answers by multiplying them together.

$$6xy$$

Here is the problem and the answer.

$$6x \cdot y = 6xy$$

Division

The division sign, ÷, means divide. The expression 6 ÷ 6 is read as "six divided by six." In algebra, ÷ is seldom used to indicate division. Instead, a slash mark, /, or a horizontal fraction bar, —, is used to indicate division. For this book, we will use the horizontal fraction bar instead of the slash mark.

6/6 or $\frac{6}{6}$ means "six divided by six."

a/3 or $\frac{a}{3}$ means "a divided by three."

MATH TALK!

Watch how these mathematical expressions are changed from Math Talk into Plain English

$$3 \div 3$$
three divided by three

$$5/3$$
five divided by three

$$a/7$$
a divided by seven

You can divide like and unlike terms in algebra.

You can divide any two numbers.
$$3 \text{ divided by } 4 = \frac{3}{4}$$
$$12 \text{ divided by } 6 = \frac{12}{6} = 2$$

You can divide any two of the same variables.
$$x \text{ divided by } x = \frac{x}{x} = 1$$

You can also divide any two different variables together.
$$a \text{ divided by } b = \frac{a}{b}$$
$$x \text{ divided by } y = \frac{x}{y}$$

You can divide two expressions that have both numbers and variables multiplied together. In order to divide these expressions it takes three steps.

1. Divide the coefficients.

2. Divide the variables.

3. Multiply both answers together.

Here are a couple of examples. Watch. It's *painless*.

Divide 3x by 4x.
First divide the coefficients.
$$3 \text{ divided by } 4 = \frac{3}{4}$$
Next divide the variables.
$$x \text{ divided by } x = \frac{x}{x} = 1$$
Finally, multiply both answers together.
$$\frac{3}{4} \text{ times } 1 \text{ is } \frac{3}{4}, \text{ so the answer is } \frac{3}{4}.$$

Here is the problem and the answer.
$$\frac{3x}{4x} = \frac{3}{4}$$

Divide 8*x* by 2*y*.

First divide the coefficients.

$$8 \text{ divided by } 2 = \frac{8}{2} = 4$$

Next divide the variables.

$$x \text{ divided by } y = \frac{x}{y}$$

Multiply both answers together.

$$4\,\frac{x}{y}$$

Here is the problem and the answer.

$$\frac{8x}{2y} = \left(\frac{4}{1}\right)\left(\frac{x}{y}\right) = \frac{4x}{y}$$

Divide 12*xy* by *x*.

First divide the coefficients. The coefficients are 12 and 1.

$$12 \text{ divided by } 1 = \frac{12}{1} = 12$$

Next divide the variables.

$$xy \text{ divided by } x = \frac{xy}{x} = y, \text{ since } \frac{x}{x} = 1$$

Finally, multiply both answers together.

$$12y$$

Here is the problem and the answer.

$$\frac{12xy}{x} = 12y$$

Divide 4*y* by 8*xy*.

First divide the coefficients. The coefficients are 4 and 8.

$$4 \text{ divided by } 8 = \frac{4}{8} = \frac{1}{2}$$

Next divide the variables.

$$y \text{ divided by } xy = \frac{y}{xy} = \frac{1}{x}, \text{ since } \frac{y}{y} = 1$$

Finally, multiply both answers together.

$$\left(\frac{1}{2}\right)\left(\frac{1}{x}\right) = \frac{1}{2x}$$

Here is the problem and the answer.

$$\frac{4y}{8xy} = \frac{1}{2x}$$

BRAIN TICKLERS
Set # 2

Solve these multiplication and division problems.

1. $3x$ times $4y$

2. $6x$ times $2x$

3. $2x$ times 5

4. $7x$ divided by $7x$

5. $4xy$ divided by $2x$

6. $3x$ divided by 3

7. $8xy$ divided by y

(Answers are on page 32.)

ZERO

Zero is an unusual number. It is neither positive nor negative. There are some rules about zero you should know. If zero is added to any number or variable, the answer is still that number or variable.

$$5 + 0 = 5$$
$$12 + 0 = 12$$
$$x + 0 = x$$

If any number or variable is added to zero, the answer is still that number or variable.

$$0 + 9 = 9$$
$$0 + \frac{1}{2} = \frac{1}{2}$$
$$0 + a = a$$

If you subtract zero from any number or variable, the answer is still that number or variable.

$$3 - 0 = 3$$
$$\frac{1}{4} - 0 = \frac{1}{4}$$
$$b - 0 = b$$

If you subtract a number or variable from zero, the answer is the opposite of that number or variable.

$$0 - 3 = -3$$
$$0 - (-4) = 4$$
$$0 - b = -b$$

If you multiply any number or variable by 0, the answer is always 0.

$$3 \cdot 0 = 0$$
$$1000(0) = 0$$
$$a \cdot 0 = 0$$
$$d(0) = 0$$
$$7xy \cdot 0 = 0$$

If you multiply 0 by any number or variable, the answer is always 0.

$$0 \cdot 7 = 0$$
$$0 \cdot 9 = 0$$
$$0(x) = 0$$

If zero is divided by any number or variable, the answer is always zero.

$$0 \div 3 = 0$$
$$0 \div (-5) = 0$$
$$\frac{0}{f} = 0$$

You cannot divide by zero. Division by zero is undefined.

$3 \div 0 = ?$ Division by zero is undefined.
$\frac{a}{0} = ?$ Division by zero is undefined.

Caution—Major Mistake Territory!

You can never divide by zero. Division by zero is undefined. Don't get trapped into thinking $5 \div 0$ is 0 or $5 \div 0$ is 5; $5 \div 0$ is undefined.

BRAIN TICKLERS
Set # 3

All of these problems have zero in them. Solve them.

1. $0 + a$

2. $a(0)$

3. $0 - a$

4. $\frac{0}{a}$

5. $(0)a$

6. $a - 0$

7. $\frac{a}{0}$

(Answers are on page 33.)

ORDER OF OPERATIONS

When you solve a mathematical sentence or expression, it's important that you do things in the correct order. The order in which you solve a problem may affect the answer.

Look at the following problem

$$3 + 1 \cdot 6$$

You read the problem as "three plus one times six," but is it "the quantity three plus one times six," which is written as $(3 + 1) \times 6$, or is it "three plus the quantity one times six," which is written as $3 + (1 \times 6)$?

These two problems have different answers.

$$(3 + 1) \times 6 = 24$$
$$3 + (1 \times 6) = 9$$

Which is the correct answer?

Mathematicians have agreed on a certain order to be used in solving mathematical problems called the "Order of Operations." The Order of Operations was developed to avoid confusion. Without the Order of Operations, several different answers would be possible when computing mathematical expressions. The Order of Operations tells you how to simplify any mathematical expression in four easy steps.

Step 1: Do everything in parentheses.
In the problem $7(6 - 1)$, subtract first, then multiply.

Step 2: Compute the value of any exponential expressions.
In the problem $5 \cdot 3^2$, square the three first and then multiply by five.

Step 3: Multiply and/or divide. Start at the left and go to the right.
In the problem $5 \cdot 2 - 4 \cdot 3$, multiply five times two and then multiply four times three. Subtract last.

Step 4: Add and/or subtract. Start on the left side of the equation and go to the right side of the equation.
In the problem $6 - 2 + 3 - 4$, start with six, subtract two, add three, and subtract four.

To remember the Order of Operations just remember the sentence, "**P**lease **E**xcuse **M**y **D**ear **A**unt **S**ally!" The first letter of each of the words tells you what to do next. The "P" in Please stands for parentheses. The "E" in Excuse stands for exponents.

The "M" in My stands for multiply. The "D" in Dear stands for divide. The "A" in Aunt stands for add. The "S" in Sally stands for subtract. If you don't forget your Aunt Sally you'll never forget the Order of Operations.

MATH TALK!

Watch as these mathematical expressions are changed from Math Talk into Plain English

$$3(5 + 2) - 7$$

three times the quantity five plus two, then that quantity minus seven

$$3 - 7(7) - 2$$

three minus the quantity seven times seven, then that quantity minus two

$$6 - 3^2 + 2$$

six minus the quantity three squared, then that quantity plus two

Watch how the following expression is evaluated.

Compute the value of $3(5 - 2) + 6 \cdot 1$

Step 1: Do what is inside the parentheses.
$5 - 2$ is inside the parentheses.
$5 - 2 = 3$
Substitute (3) for $(5 - 2)$.
$3(3) + 6 \cdot 1$

Step 2: Compute the value of any exponential expressions.
There are no exponential expressions.
Go to the next step.

Step 3: Multiply and/or divide from left to right.
Multiply three times three and then six times one.
$(3)(3) = 9$ and $(6)(1) = 6$
Substitute 9 for $(3)(3)$ and 6 for $(6)(1)$.
$9 + 6$

Step 4: Add and/or subtract from left to right.
Add $9 + 6$.
$9 + 6 = 15$
Solution: $3(5 - 2) + 6 \cdot 1 = 15$

Watch how the following expression is evaluated.

$$(4 - 1)^2 - 2 \cdot 3$$

Step 1: Do what is inside the parentheses first.
$4 - 1$ is inside the parentheses.
$4 - 1 = 3$
Substitute 3 for $4 - 1$.
$(3)^2 - 2 \cdot 3$

Step 2: Compute the value of any exponential expressions.
Square 3.
$3^2 = 9$
Substitute 9 for 3^2.
$9 - 2 \cdot 3$

Step 3: Multiply and/or divide from left to right.
Multiply $2 \cdot 3$.
$2 \cdot 3 = 6$
Substitute 6 for $2 \cdot 3$.
$9 - 6$

Step 4: Add and/or subtract from left to right.
Subtract 6 from 9.
$9 - 6 = 3$
Solution: $(4 - 1)^2 - 2 \cdot 3 = 3$

SUM IT UP!

When evaluating expressions, remember
"Please Excuse My Dear Aunt Sally."
Parentheses
Exponents
Multiplication
Division
Addition
Subtraction

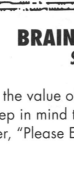

BRAIN TICKLERS
Set # 4

Compute the value of the following expressions. Keep in mind the Order of Operations. Remember, "Please Excuse My Dear Aunt Sally."

1. $4 \div (2 + 2) - 1$

2. $3 + 12 - 5 \cdot 2$

3. $16 - 2 \cdot 4 + 3$

4. $6 + 5^2 - 12 + 4$

5. $(4 - 3)^2(2) - 1$

(Answers are on page 33.)

PROPERTIES OF NUMBERS

There are five properties of numbers that are important in the study of algebra.

> The Commutative Property of Addition
> The Commutative Property of Multiplication
> The Associative Property of Addition
> The Associative Property of Multiplication
> The Distributive Property of Multiplication over Addition

What do each of these properties say?

The commutative property of addition

The Commutative Property of Addition states that no matter in what order you add two numbers, the sum is always the same. In other words, three plus four is equal to four plus three. Six plus two is equal to two plus six.

Written in Math Talk, the Commutative Property of Addition is $a + b = b + a$. Given any two numbers a and b, a plus b is equal to b plus a.

EXAMPLES:

$3 + 5 = 5 + 3$ because $8 = 8$.

$\frac{1}{2} + 6 = 6 + \frac{1}{2}$ because $6\frac{1}{2} = 6\frac{1}{2}$.

$5x + 3 = 3 + 5x$

Caution—Major Mistake Territory!

Subtraction is not commutative. The order of the numbers in a subtraction problem *does* make a difference.

$6 - 3$ is not the same as $3 - 6$.
$5 - 0$ is not the same as $0 - 5$.

The commutative property of multiplication

The Commutative Property of Multiplication states that no matter in what order you multiply two numbers, the answer is always the same.

Written in Math Talk! The Commutative Property of Multiplication is $a(b) = b(a)$. Given any two numbers, a and b, a times b is equal to b times a.

EXAMPLES:

$3 \cdot 5 = 5 \cdot 3$ because $15 = 15$.

$6(1) = 1(6)$ because $6 = 6$.

$\frac{1}{2}(4) = 4\left(\frac{1}{2}\right)$ because $2 = 2$.

Caution—Major Mistake Territory!

Division is not commutative. The order of the numbers in a division problem *does* make a difference.

$5 \div 10$ is not the same as $10 \div 5$.

$\frac{6}{2}$ is not the same as $\frac{2}{6}$.

$\frac{a}{2}$ is not the same as $\frac{2}{a}$.

The associative property of addition

The Associative Property of Addition states that when you add three numbers together, no matter how you group them, the answer is still the same.

Written in Math Talk, the Associative Property of Addition is $(a + b) + c = a + (b + c)$. If you first add a and b and then add c to the total, the answer is the same as if you first add b and c and then add the total to a.

EXAMPLES:

$(3 + 5) + 2 = 3 + (5 + 2)$ because $8 + 2 = 3 + 7$.
$(1 + 8) + 4 = 1 + (8 + 4)$ because $9 + 4 = 1 + 12$.

The associative property of multiplication

The Associative Property of Multiplication states that when you multiply three numbers together, no matter how you group them, the product is always the same.

Written in Math Talk, the Associative Property of Multiplication is $(a \cdot b)c = a(b \cdot c)$. If you first multiply a and b and then multiply the product by c, it is the same as if you first multiply b and c and then multiply a by the product.

EXAMPLES:

$(3 \cdot 2)6 = 3(2 \cdot 6)$ because $(6)6 = 3(12)$.
$(5 \cdot 4)2 = 5(4 \cdot 2)$ because $(20)2 = 5(8)$.

The distributive property of multiplication over addition

The Distributive Property of Multiplication over Addition states that when you multiply a monomial expression such as 3 times a binomial expression such as $(2 + x)$, the answer is the monomial (3) times the first term of the binomial expression (2) plus the monomial (3) times the second term of the binomial expression (x).

Written in Math Talk, the Distributive Property of Multiplication over Addition is $a(b + c) = ab + ac$. Multiplying a times the quantity b plus c is equal to a times b plus a times c.

EXAMPLES:

$3(5 + 2) = 3 \cdot 5 + 3 \cdot 2$ because $3(7) = 15 + 6$.

$\frac{1}{2}(4 + 1) = \frac{1}{2}(4) + \frac{1}{2}(1)$ because $\frac{1}{2}(5) = 2 + \frac{1}{2}$.

$6(3 + x) = 6(3) + 6(x)$ or $6(3 + x) = 18 + 6x$.

$5y(2x + 3) = 5y(2x) + 5y(3)$ or $5y(2x + 3) = 10xy + 15y$.

BRAIN TICKLERS
Set # 5

Use the following abbreviations.

CA = The Commutative Property of Addition
CM = The Commutative Property of Multiplication
AA = The Associative Property of Addition
AM = The Associative Property of Multiplication
DM/A = The Distributive Property of Multiplication over Addition

Next to each mathematical equation, write the abbreviation for the property the equation represents. Be careful—some of the problems are tricky.

____ 1. $6(5 + 1) = 6(5) + 6(1)$

____ 2. $4 + (3 + 2) = (4 + 3) + 2$

____ 3. $5 + 3 = 3 + 5$

____ 4. $3(5 \cdot 1) = (3 \cdot 5)1$

____ 5. $7(3) = 3(7)$

____ 6. $6(4 + 3) = 6(3 + 4)$

(Answers are on page 33.)

NUMBER SYSTEMS

There are six different number systems.

The Natural Numbers
The Whole Numbers
The Integers
The Rational Numbers
The Irrational Numbers
The Real Numbers

The natural numbers

The natural numbers are the numbers 1, 2, 3, 4, 5, ...
The three dots . . . mean to continue counting forever.

The natural numbers are sometimes called the *counting numbers* because they are the numbers that you use to count.

EXAMPLES:

7 and 9 are natural numbers.

$0, \frac{1}{2}$, and -3 are *not* natural numbers.

The whole numbers

The whole numbers are the numbers 0, 1, 2, 3, 4, 5, 6, . . .
The whole numbers are the natural numbers plus zero.
All of the natural numbers are whole numbers.

EXAMPLES:

0, 5, 23, and 1001 are whole numbers.

$-4, \frac{1}{3}$, and 0.2 are *not* whole numbers.

The integers

The integers are the natural numbers, their opposites, and
   zero.
The integers are . . . , $-3, -2, -1, 0, 1, 2, 3, ...$
All of the whole numbers are integers.
All of the natural numbers are integers.

EXAMPLES:

$-62, -12, 27$, and 83 are integers.

$-\frac{3}{4}, \frac{1}{2}$, and $\sqrt{2}$ are *not* integers.

The rational numbers

The rational numbers are any numbers that can be expressed
   as the ratio of two whole numbers.
All of the integers are rational numbers.
All of the whole numbers are rational numbers.
All of the natural numbers are rational numbers.

EXAMPLES:

3 can be written as $\frac{3}{1}$, so 3 is a rational number.

-27, $-12\frac{1}{2}$, $-\frac{1}{3}$, $\frac{1}{4}$, 7, $4\frac{4}{5}$, and 1,000,000 are all rational numbers.

$\sqrt{2}$ and $\sqrt{3}$ are *not* rational numbers.

The irrational numbers

The irrational numbers are numbers that cannot be expressed as the ratio of two whole numbers.
The rational numbers are not irrational numbers.
The integers are not irrational numbers.
The whole numbers are not irrational numbers.
The natural numbers are not irrational numbers.

EXAMPLES:

$-\sqrt{2}$, $\sqrt{2}$, and $\sqrt{3}$ are irrational numbers.

-41, $-17\frac{1}{2}$, $-\frac{3}{8}$, $\frac{1}{5}$, 4, $\frac{41}{7}$, and 1,247 are *not* irrational numbers.

The real numbers

The real numbers are a combination of all the number systems.
The real numbers are the natural numbers, whole numbers, integers, rational numbers, and the irrational numbers.
Every point on the number line is a real number.
All of the irrational numbers are real numbers.
All of the rational numbers are real numbers.
All of the integers are real numbers.
All of the whole numbers are real numbers.
All of the natural numbers are real numbers.

EXAMPLES:

-53, $-\frac{17}{3}$, $4\frac{1}{2}$, $-\sqrt{2}$, $-\frac{3}{5}$, 0, $\frac{1}{6}$, $\sqrt{3}$, 4, $\frac{41}{7}$ and 1,247 are all real numbers.

Caution—Major Mistake Territory!

Is five a whole number or a natural number? Five is both a whole number and a natural number. A number can belong to more than one number system at the same time.

Is six a whole number or a rational number? Six is both a whole number and a rational number. Six can be written as 6 or as $\frac{6}{1}$.

BRAIN TICKLERS
Set # 6

Using the abbreviations below, circle the number systems each number belongs to. Some numbers belong to more than one number system.

N = Natural Numbers
W = Whole Numbers
In = Integers
Ra = Rational Numbers
Ir = Irrational Numbers
Re = Real Numbers

1. 3	N	W	In	Ra	Ir	Re
2. -7	N	W	In	Ra	Ir	Re
3. 0	N	W	In	Ra	Ir	Re
4. $-\frac{1}{4}$	N	W	In	Ra	Ir	Re
5. $\frac{3}{8}$	N	W	In	Ra	Ir	Re
6. $\frac{6}{1}$	N	W	In	Ra	Ir	Re
7. -4	N	W	In	Ra	Ir	Re
8. 6	N	W	In	Ra	Ir	Re
9. $\sqrt{3}$	N	W	In	Ra	Ir	Re
10. $\frac{12}{3}$	N	W	In	Ra	Ir	Re

(Answers are on page 34.)

BRAIN TICKLERS—
THE ANSWERS

Set # 1, page 9

1. $3x + 7x = 10x$

2. $4x + x = 5x$

3. $3x - 3x = 0$

4. $10x - x = 9x$

5. $6x - 4x = 2x$

6. $3x + 2 = 3x + 2$ The sum of these terms cannot be simplified. They are not like terms.

7. $10 - 4x = 10 - 4x$ This expression cannot be simplified because 10 and $4x$ are not like terms.

Set # 2, page 16

1. $3x$ times $4y = 3x(4y) = 12xy$

2. $6x$ times $2x = 6x(2x) = 12x^2$

3. $2x$ times $5 = 2x(5) = 10x$

4. $7x$ divided by $7x = \frac{7x}{7x} = 1$

5. $4xy$ divided by $2x = \frac{4xy}{2x} = 2y$

6. $3x$ divided by $3 = \frac{3x}{3} = x$

7. $8xy$ divided by $y = \frac{8xy}{y} = 8x$

Set # 3, page 18

1. $0 + a = a$

2. $a(0) = 0$

3. $0 - a = -a$

4. $\frac{0}{a} = 0$

5. $(0)a = 0$

6. $a - 0 = a$

7. $\frac{a}{0}$ is undefined.

Set # 4, page 23

1. $4 \div (2 + 2) - 1 = \frac{4}{4} - 1 = 1 - 1 = 0$

2. $3 + 12 - 5 \cdot 2 = 3 + 12 - 10 = 15 - 10 = 5$

3. $16 - 2 \cdot 4 + 3 = 16 - 8 + 3 = 11$

4. $6 + 5^2 - 12 + 4 = 6 + 25 - 12 + 4 = 23$

5. $(4 - 3)^2(2) - 1 = (1)^2(2) - 1 = 1(2) - 1 = 2 - 1 = 1$

Set # 5, page 27

1. DM/A

2. AA

3. CA

4. AM

5. CM

6. CA

Set # 6, page 31

1. N, W, In, Ra, Re

2. In, Ra, Re

3. W, In, Ra, Re

4. Ra, Re

5. Ra, Re

6. N, W, In, Ra, Re

7. In, Ra, Re

8. N, W, In, Ra, Re

9. Ir, Re

10. N, W, In, Ra, Re

CHAPTER TWO

The Integers

The weather forecaster for the preceding report used negative numbers to express how cold it was outside, because the temperature was below zero. When you first learned to count, you counted with positive numbers, the numbers greater than zero: 1, 2, 3, 4, 5. That's why these numbers are called the counting numbers. You probably first learned to count to ten, then to 100, and maybe even to 1,000. Finally, you learned that you could continue counting forever, because after every number there is a number that is one larger than the number before it.

Well, there are negative numbers, too. Negative numbers are used to express cold temperatures, money owed, feet below sea level, and lots of other things. You count from negative one to negative ten with these numbers: −1, −2, −3, −4, −5, −6, −7, −8, −9, −10. You could continue counting until you reached −100 or −1,000, or you could continue counting forever. After every negative number there is another negative number that is less than the number before it.

When you place all these positive numbers, negative numbers, and zero together, you have what mathematicians call the *integers*.

WHAT ARE THE INTEGERS?

The integers are made up of three groups of numbers:

- the positive integers
- the negative integers
- zero

The positive integers are:
$$1, 2, 3, 4, \ldots$$

MATH TALK!

Three dots together, . . . , means to continue the pattern forever.

Sometimes the positive integers are written like this:
$$+1, +2, +3, +4, \ldots$$

Or, sometimes they are written like this:
$$(+1), (+2), (+3), (+4), \ldots$$

Here is a graph of the positive integers:

Notice that the positive integers are only the counting numbers. The numbers between any two counting numbers are not integers. The heavy arrow pointing to the right means to continue counting forever in that direction.

MATH TALK!

If there is no sign in front of a number,
you can assume it is positive.

4 can also be written as $+4$ or $(+4)$.
$+4$ is read as "positive 4."
$+4$ is not read as "plus 4."

The negative integers are
$$-1, -2, -3, -4, -5, \ldots$$

Sometimes they are written like this:
$$(-1), (-2), (-3), (-4), (-5) \ldots$$

Here is a graph of the negative integers:

$$-10 \quad -9 \quad -8 \quad -7 \quad -6 \quad -5 \quad -4 \quad -3 \quad -2 \quad -1$$

MATH TALK!

-3 can also be written as (-3).
-3 is read as "negative three."
-3 is not read as "minus three."

Zero is an integer, but it is neither positive nor negative.

Here is a graph of zero.

Here is a list representing all the integers.

$$\ldots -4, -3, -2, -1, 0, 1, 2, 3, 4 \ldots.$$

Here is a graph representing all the integers.

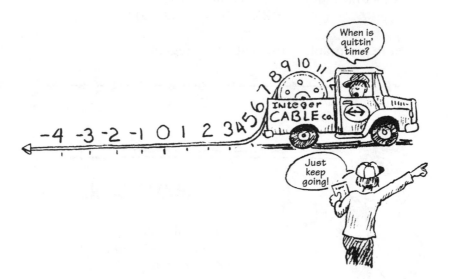

Here is a list of some numbers that are integers.

$$0, 4, -7, -1000, 365, \frac{10}{2}, -\frac{6}{3}, \frac{10}{10}, 123{,}456{,}789$$

Here is list of some numbers that are not integers.

$$7.2, \frac{6}{4}, -\frac{3}{8}, -1.2$$

Caution—Major Mistake Territory!

The number $\frac{6}{2}$ is an integer. The number $\frac{5}{2}$ is *not* an integer. Why? The number $\frac{6}{2}$ is 6 divided by 2, which is 3, and 3 is an integer. The number $\frac{5}{2}$ is 5 divided by 2, which is $2\frac{1}{2}$, and $2\frac{1}{2}$ is not an integer.

Graphing integers

Here's how different integers look when they are graphed on the number line.

Integer: (-2); Graph:

Integer: (-1); Graph:

Integer: 3; Graph:

Integer: (-4); Graph:

Integer: 0; Graph:

WHICH IS GREATER?

Sometimes mathematicians want to compare two numbers and decide which is larger and which is smaller. But instead of *larger* and *smaller*, mathematicians use the words *greater than* and *less than*. They say, "Seven is *greater than* three." or "Three is *less than* seven." Mathematicians use the symbols ">" or "<" to indicate these relationships.

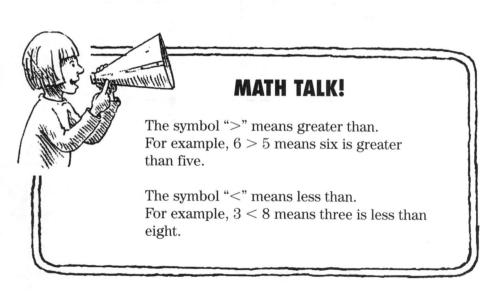

MATH TALK!

The symbol ">" means greater than.
For example, 6 > 5 means six is greater than five.

The symbol "<" means less than.
For example, 3 < 8 means three is less than eight.

Here is an example of two inequalities that mean the same thing. Notice that the arrow always points to the smaller number.

$2 > 1$ *(Two is greater than one.)*
$1 < 2$ *(One is less than two.)*

MATH TALK!

Mathematical sentences with ">" or "<" are called *inequalities*.

Mathematical sentences with "=" are called *equalities* or *equations*.

Now let's look at some pairs of positive and negative integers and decide which is larger and which is smaller.

- -3 and 5
  Hint: Positive numbers are always greater than negative numbers.
  $-3 < 5$ (Negative three is less than five.)
  Or, you could write
  $5 > -3$ (Positive five is greater than negative three.)

- 6 and 0
  Hint: Positive numbers are always greater than zero.
  $6 > 0$ (Six is greater than zero.)
  Or, you could write
  $0 < 6$ (Zero is less than six.)

- -3 and 0
  Hint: Negative numbers are always less than zero.
  $-3 < 0$ (Negative three is less than zero.)
  Or, you could write
  $0 > -3$ (Zero is greater than negative three.)

- -4 and -10

  Hint: The larger a negative number looks, the smaller it actually is.

  $-4 > -10$ (Negative four is greater than negative ten.)
  Or, you could write

  $-10 < -4$ (Negative ten is less than negative four.)

Caution—Major Mistake Territory!

When looking at signed numbers it's easy to get confused about which number is larger.

Remember:
1. Positive numbers are always greater than negative numbers.
2. Zero is always less than a positive number.
3. Zero is always greater than a negative number.
4. When comparing two negative numbers, the number that looks larger is actually smaller. Negative six looks larger than negative two, but it is actually smaller. Still confused? Graph both numbers on a number line. The negative number closest to zero is always larger.

BRAIN TICKLERS
Set # 7

Mark each statement true (T) or false (F).

Hint: Five of these statements are true. The sum of the problem numbers of the true statements is 25.

_____ 1. -7 is an integer.

_____ 2. $\frac{3}{2}$ is an integer.

_____ 3. 5 is an integer.

_____ 4. 0 is a positive integer.

_____ 5. $2 < 10$

_____ 6. $(-5) > (8)$

_____ 7. $(-7) < (-5)$

_____ 8. $(-3) < (-5)$

_____ 9. $(-1) > (-4)$

_____10. $0 < -1$

(Answers are on page 65.)

ADDING INTEGERS

When you add integers, the problem may look like one of the four types of problems below.

Case 1: Both numbers are positive.
   Example: $(+3) + (+5)$

Case 2: Both numbers are negative.
   Example: $(-2) + (-4)$

Case 3: One of the two numbers is positive and one is negative.
   Example: $(+3) + (-8)$

Case 4: One of the two numbers is zero.
   Example: $0 + (+2)$

Now let's see how to solve each of these types of problems.

Case 1: Both numbers are positive.

Painless Solution: Add the numbers just like you would add any two numbers. The answer is always positive.

Take this
Painless Solution
and call me
in the morning.

EXAMPLES:

$(+3) + (+8) = +11$
$2 + (+4) = +6$
$3 + 2 = 5$

Case 2: Both numbers are negative.

Painless Solution: Pretend both numbers are positive. Add them. Place a negative sign in front of the answer.

EXAMPLES:

$(-3) + (-8) = (-11)$
$(-2) + (-4) = (-6)$
$-5 + (-5) = -10$

Case 3: One number is positive and one number is negative.

Painless Solution: Pretend both numbers are positive. Subtract the smaller number from the larger number. Give the answer the sign of the number that would be larger if both numbers were positive.

$(-3) + (+8) = ?$
Pretend both numbers are positive and subtract the smaller number from the larger number.
$$8 - 3 = 5$$
Give the answer the sign of the larger number if both numbers were positive. Eight is larger than three and it is positive, so the answer is positive.
$$(-3) + (+8) = +5$$

$(+2) + (-4) = ?$
Pretend both numbers are positive and subtract the smaller number from the larger number.
$$4 - 2 = 2$$
Give the answer the sign of the larger number if both numbers were positive. Four is larger than two and it is negative, so the answer is negative.
$$(+2) + (-4) = -2$$

Case 4: One of the numbers is zero.

Painless Solution: Zero plus any number is that number.

EXAMPLES:

$$(+2) + 0 = +2$$
$$0 + (-8) = -8$$

MATH TALK!

The numbers you add in an addition problem are called *addends*.

The answer of an addition problem is called the *sum*.

In the problem, $(3) + (-2) = 1$,

3 is an addend,
(-2) is an addend,
1 is the sum.

BRAIN TICKLERS
Set # 8

Solve the following addition problems.

____ 1. $(+5) + 0$

____ 2. $(+3) + (+6)$

____ 3. $(-3) + (+6)$

____ 4. $0 + (-1)$

____ 5. $(+4) + (-4)$

____ 6. $(-6) + (-3)$

____ 7. $(-5) + (+2)$

____ 8. $5 + 2$

____ 9. $2 + (-4)$

____10. $(-5) + (-2)$

(Answers are on page 65.)

SUBTRACTING INTEGERS

When you subtract one integer from another integer, there are six possible cases.

Case 1: Both numbers are positive.
Example: $5 - (+3)$

Case 2: Both numbers are negative.
Example: $(-7) - (-4)$

Case 3: The first number is positive and the second is negative.
Example: $3 - (-4)$

Case 4: The first number is negative and the second number is positive.
Example: $(-5) - (+3)$

Case 5: The first number is zero.
Example: $0 - (-3)$

Case 6: The second number is zero.
Example: $3 - (0)$

MATH TALK!

The first number in a subtraction problem is called the *minuend*.

The second number in a subtraction problem is called the *subtrahend*.

The answer in a subtraction problem is called the *difference*.

Wow! Six possible cases. How do you remember how to subtract one number from another? It's easy with the *Painless Solution*.

Change the subtraction problem into an addition problem by taking the opposite of the number being subtracted. Now, solve the problem just like any other addition problem.

MATH TALK!

To take the opposite of a number, reverse the sign.

The opposite of 7 is (-7).
The opposite of 242 is (-242).
The opposite of -13 is 13.
The opposite of (-6) is 6.
The opposite of 0 is always 0.

Watch how easy it is to solve these subtraction problems with the *Painless Solution*.

$(+7) - (-3)$
Change the subtraction problem to an addition problem.
$$(+7) + (-3)$$
Take the opposite of the number being subtracted.
$$(+7) + (+3)$$
Solve the problem.
$$(+7) + (+3) = 10$$

$(-4) - (-3)$
Change the subtraction problem to an addition problem.
$$(-4) + (-3)$$
Take the opposite of the number being subtracted.
$$(-4) + (+3)$$
Solve the problem.
$$(-4) + (+3) = -1$$

$$0 - (+5)$$

Change the subtraction problem to an addition problem.

$$0 + (+5)$$

Take the opposite of the number being subtracted.

$$0 + (-5)$$

Solve the problem.

$$0 + (-5) = -5$$

Caution—Major Mistake Territory!

Make sure you change both signs, not just one.

Change the sign of the problem and the sign of the number being subtracted.

BRAIN TICKLERS
Set # 9

The first step in solving a subtraction problem is knowing how to change a subtraction problem into an addition problem. Match the subtraction problems to the correct addition problems. If you get them all correct, you will spell a word.

___	6 − (−4)	T.	6 + (4)
___	−6 − (4)	I.	6 + (−4)
___	6 − (4)	C.	−6 + (4)
___	(−6) − (−4)	R.	−6 + (−4)
___	0 − (−4)	Y.	0 + (−4)
___	0 − (+4)	K.	0 + 4

(Answers are on page 66.)

BRAIN TICKLERS
Set # 10

Now solve these same subtraction problems. Use the addition problems in Brain Ticklers #3 for help.

____ 1. $6 - (-4)$

____ 2. $-6 - (4)$

____ 3. $6 - (4)$

____ 4. $(-6) - (-4)$

____ 5. $0 - (-4)$

____ 6. $0 - (+4)$

(Answers are on page 66.)

Remember . . .
Just change the subtraction problem into an addition problem and take the opposite of the number being subtracted. Solve the new addition problem.

Example: $6 - (-4)$
$6 - (-4)$ becomes $6 + (+4)$ after changing both signs.
$6 + (4) = 10$

MULTIPLYING INTEGERS

Multiplying integers is easy. When you multiply two integers together, there are four possible cases.

Case 1: Both numbers are positive.
Example: $6 \cdot 4$

Case 2: Both numbers are negative.
Example: $(-3)(-2)$

Case 3: One number is positive and the other is negative.
Example: $(-5)(+2)$

Case 4: One of the two numbers is zero.
Example: $(-6) \cdot 0$

Here is how you solve multiplication problems with integers.

Case 1: Both numbers are positive.

Painless Solution: Just multiply the numbers. The answer is always positive.

EXAMPLES:

$5 \times 3 = 15$
$(+6)(+4) = (+24)$
$2 \times (+7) = (+14)$

Case 2: Both numbers are negative.

Painless Solution: Just pretend the numbers are positive. Multiply the numbers. The answer is always positive.

EXAMPLES:

$(-5)(-3) = 15$
$-6 \times (-4) = +24$

Case 3: One number is positive and the other negative.

Painless Solution: Just pretend the numbers are positive. Multiply the numbers together. The answer is always negative.

EXAMPLES:

$$(-4)(+3) = (-12)$$
$$5 \times (-2) = (-10)$$

Case 4: One of the two numbers is zero.

Painless Solution: The answer is always 0. It doesn't matter if you are multiplying a positive number by zero or a negative number by zero. Zero times any number and any number times zero is always zero.

EXAMPLES:

$$0 \times 7 = 0$$
$$(-8) \times 0 = 0$$
$$(+4) \times 0 = 0$$
$$0 \times (-1) = 0$$

MATH TALK!

The numbers you multiply together in a multiplication problem are called *factors*.

The answer in a multiplication problem is called the *product*.

In the problem $(5)(-6) = (-30)$,
5 and (-6) are factors,
(-30) is the product.

Remember . . .

A positive number times a positive number is a positive number.

A negative number times a negative number is a positive number.

A positive number times a negative number is a negative number.

BRAIN TICKLERS
Set # 11

Solve the following multiplication problems.

1. $(-2)(-8)$

2. $(3)(-3)$

3. $(8)(-2)$

4. $5 \cdot 0$

5. $3 \cdot 3$

(Answers are on page 66.)

MATH TALK!

The first number in a division problem is called the *dividend*.

The second number in a division problem is called the *divisor*.

The answer in a division problem is called the *quotient*.

In the division problem $(-8) \div 2 = (-4)$,
-8 is the dividend,
2 is the divisor,
-4 is the quotient.

DIVIDING INTEGERS

When you divide two integers, there are five possible cases.

Case 1: Both numbers are positive.
Example: $21 \div 7$

Case 2: Both numbers are negative.
Example: $(-15) \div (-3)$

Case 3: One number is negative and one number is positive.
Example: $(+8) \div (-4)$

Case 4: The dividend is 0.
Example: $0 \div (-2)$

Case 5: The divisor is 0.
Example: $6 \div 0$

Here is how you solve each of these cases.

Case 1: Both numbers are positive.

Painless Solution: Divide the numbers. The answer is always positive.

EXAMPLES:

$8 \div 2 = 4$

$(+12) \div (+4) = 3$

Case 2: Both numbers are negative.

Painless Solution: Pretend both numbers are positive. Divide the numbers. The answer is always positive.

EXAMPLES:

$(-9) \div (-3) = 3$

$-15 \div (-3) = +5$

MATH TALK!

Here's how you change Math Talk into just Plain English.

$(-9) \div (-3) = 3$

Negative nine divided by negative three equals three.

Case 3: One number is positive and the other negative.

Painless Solution: Pretend the numbers are positive. Divide the numbers. The answer is always negative.

EXAMPLES:

$-9 \div 3 = (-3)$

$15 \div (-3) = -5$

Case 4: The dividend is zero.

Painless Solution: Zero divided by any number (except 0) is zero. It doesn't matter if you are dividing zero by a positive or negative number. The answer is always zero.

EXAMPLES:

$0 \div 6 = 0$

$0 \div (-3) = 0$

Case 5: The divisor is zero.

Painless Solution: Division by zero is always undefined. How can you divide something into zero parts?

EXAMPLES:

$4 \div 0 = $ undefined

$(-8) \div 0 = $ undefined

Remember . . .
The rules for division are the same as for multiplication.

A positive number divided by a positive number is positive.

A negative number divided by a negative number is positive.

A positive number divided by a negative number is negative.

A negative number divided by a positive number is negative.

BRAIN TICKLERS
Set # 12

Solve these division problems.

1. $5 \div 5$ 4. $0 \div 10$

2. $5 \div (-5)$ 5. $(-6) \div (-3)$

3. $(-6) \div 3$ 6. $10 \div 0$

(Answers are on page 67.)

WHAT SIGN IS THE ANSWER?

When you add, subtract, multiply, or divide signed numbers, how do you figure out the sign of the answer? This chart summarizes everything you've learned. Decide the type of problem. Is it an addition, subtraction, multiplication, or division problem?

What is the sign of the problem? Now you know the sign of the answer.

	BOTH NUMBERS (+)	BOTH NUMBERS (−)	ONE NUMBER (+) & ONE NUMBER (−)
ADDITION	Positive (+)	Negative (−)	(+) or (−)
SUBTRACTION	(+) or (−)	(+) or (−)	(+) or (−)
MULTIPLICATION	(+)	(+)	(−)
DIVISION	(+)	(+)	(−)

SUPER BRAIN TICKLERS

See if you can figure out the sign of the correct answer. Circle the letter next to the correct answer and spell a phrase.

1. $(−6) + (−2) =$ (−) E or (+) S

2. $(−12) ÷ (−3) =$ (−) O or (+) A

3. $(−6) − (−2) =$ (−) S or (+) T

4. $−7 ÷ (+1) =$ (−) Y or (+) N

5. $2 − (−2) =$ (−) I or (+) A

6. $(−3)(−2) =$ (−) T or (+) S

7. $(−4) + (−7) =$ (−) P or (+) H

8. $(−6) − (−8) =$ (−) R or (+) I

9. $4 × (−2) =$ (−) E or (+) T

(Answers are on page 67.)

WORD PROBLEMS

Many people go "Ugh!" when they hear that it's time to do word problems. They think word problems are hard, but actually they are easy. To solve a word problem all you have to do is change "Plain English" to "Math Talk." Here is how to solve a few word problems that use integers.

PROBLEM 1: An elevator went up three floors and down two floors. How much higher or lower was the elevator than when it started?

Painless Solution:
The elevator went up three floors $(+3)$.
The elevator went down two floors (-2).
The word "and" means add $(+)$.
So, the problem is $(+3) + (-2)$.
The answer is $(+3) + (-2) = (+1)$.
The elevator was one floor higher than when it started.

PROBLEM 2: Today's high temperature was six degrees. Today's low temperature was two degrees below zero. What was the change in temperature?

Painless Solution:
The high temperature was six degrees $(+6)$.
The low temperature was two degrees below zero (-2).

The phrase "change" means subtract (−).
So, the problem is (+6) − (−2).
The answer is (+6) − (−2) = +8.
There was an eight degree change in temperature.

PROBLEM 3: The temperature dropped two degrees every hour.
How many degrees did it drop in six hours?

Painless Solution:
The temperature dropped two degrees (−2).
The time that went by was six hours (+6).
Type of problem: Multiplication
So, the problem is (−2) × (+6).
The answer is (−2) × (+6) = (−12).
The temperature dropped 12 degrees.

PROBLEM 4: Bob spends $3 a day on lunch. So far this week Bob
has spent a total of $12 for lunches. For how many days has Bob
bought lunch?

Painless Solution:
The total spent for lunches was $12.
The cost of lunch for one day was $3.
Type of Problem: Division
So, the problem is $12 ÷ $3.
The answer is $12 ÷ $3 = 4.
Bob bought lunch for four days.

BRAIN TICKLERS—
THE ANSWERS

Set # 7, page 45

1. T

2. F

3. T

4. F

5. T

6. F

7. T

8. F

9. T

10. F

Notice that $1 + 3 + 5 + 7 + 9 = 25$.

Set # 8, page 49

1. $(+5) + 0 = +5$

2. $(+3) + (+6) = +9$

3. $(-3) + (+6) = +3$

4. $0 + (-1) = -1$

5. $(+4) + (-4) = 0$

6. $(-6) + (-3) = -9$

7. $(-5) + (+2) = -3$

8. $5 + 2 = +7$

9. $2 + (-4) = -2$

10. $(-5) + (-2) = -7$

Set # 9, page 53

TRICKY

Set # 10, page 54

1. $6 - (-4) = 10$

2. $-6 - (4) = -10$

3. $6 - (4) = 2$

4. $(-6) - (-4) = -2$

5. $0 - (-4) = +4$

6. $0 - (+4) = -4$

Set # 11, page 57

1. $(-2)(-8) = 16$

2. $(3)(-3) = -9$

3. $(8)(-2) = -16$

4. $5 \cdot 0 = 0$

5. $3 \cdot 3 = 9$

Set # 12, page 61

1. $5 \div 5 = 1$

2. $5 \div (-5) = -1$

3. $(-6) \div 3 = -2$

4. $0 \div 10 = 0$

5. $(-6) \div (-3) = 2$

6. $10 \div 0$ is undefined.

Super Brain Ticklers, page 62

EASY AS PIE

CHAPTER THREE

Solving Equations with One Variable

DEFINING THE TERMS

An *equation* is a mathematical sentence with an equals sign in it. A *variable* is a letter that is used to represent a number. Some equations have variables in them and some do not.

> $3 + 5 = 8$ is an equation.
>
> $3x + 1 = 4$ is an equation.
>
> $2x + 7x + 1$ is not an equation. It does not have an equals sign.
>
> $3 + 2 > 5$ is a mathematical sentence, but it is not an equation because it does not have an equals sign in it.

MATH TALK!

Think of the letter x as a mystery number. Here is how to change the following equations from Math Talk into Plain English.

$$x + 3 = 7$$
A mystery number plus three equals seven.

$$2x - 4 = 8$$
Two times a mystery number minus four equals eight.

$$\tfrac{1}{2}x = 5$$
One half of a mystery number equals five.

$$3(x + 1) = 0$$
Three times the quantity of a mystery number plus one equals zero.

One of the major goals of algebra is to figure out what this mystery number is. When you figure out the value of the mystery number and put it back in the equation, the mathematical sentence will be true.

Sometimes you can look at an equation and guess what the mystery number is.

Look at the equation $x + 1 = 2$. What do you think the mystery number is? You're right, it's one: $1 + 1 = 2$.

What if you look at the equation $y - 1 = 3$. What do you think the mystery number is? You're right, it's four: $4 - 1 = 3$.

What do you think "a" stands for in the equation $2a = 10$? You're right, it's five: $2(5) = 10$.

Here's an easy one. What is "b" in the equation $3 + 4 = b$? You're right, it's seven: $3 + 4 = 7$.

Sometimes you can look at an equation and figure out the correct answer, but most of the time you have to solve an equation using the principles of algebra. Could you solve the equation $3(x + 2) + 5 = 6(x - 1) - 4$ in your head? Probably not.

You would have to use some of the principles of algebra to solve that equation. By the time you finish this chapter you will consider this equation as easy as pie.

Easy as pie!

$3(x+2) + 5 = 6(x-1) - 4$

SOLVING EQUATIONS

Solving equations is *painless*. There are three steps to solving an equation with one variable.

Step 1: Simplify each side of the equation.

Step 2: Add and/or Subtract.

Step 3: Multiply or Divide.

Remember the three steps, Simplify, Add and/or Subtract, and Multiply or Divide, and you will be able to solve multitudes of equations with one variable. Here is how to do each step.

Step 1: Simplifying each side of the equation

To simplify an equation, first simplify the left side of the equation. Next simplify the right side of the equation. When you simplify each side of the equation, use the Order of Operations. On each side, do what is inside the parentheses first. Use the Distributive Property of Multiplication over Addition to get rid of the parentheses. Multiply and divide. Add and subtract.

$5x = 3(4 + 1)$
First simplify the left side of the equation.
There is nothing to simplify on the left side of the equation, so simplify the right side of the equation. First add what is inside the parentheses.
$$4 + 1 = 5$$
Substitute 5 for 4 + 1.
$$5x = 3(5)$$
Next, multiply 3(5).
$$5x = 15$$
Now this equation is ready to be solved.

$5(x + 2) = 10 + 5$
First simplify the left side of the equation. Since you cannot add what is inside the parentheses, multiply $5(x + 2)$ using the Distributive Property of Multiplication over Addition. Multiply 5 times x and 5 times 2.
$$5(x) + 5(2) = 5x + 10$$
Next simplify the right side of the equation.
$$10 + 5 = 15$$
Substitute 15 for 10 + 5.
$$5x + 10 = 15$$
This equation is ready to be solved.

$3x + 4x = 6(2x + 1)$
Simplify the left side of the equation. Add $3x + 4x$.
$$3x + 4x = 7x$$
Substitute $7x$ for $3x + 4x$.
$$7x = 6(2x + 1)$$
Next simplify the right side of the equation. Since you cannot add $2x + 1$, multiply $6(2x + 1)$.
$$6(2x + 1) = 12x + 6$$

Substitute $12x + 6$ for $6(2x + 1)$.
$$7x = 12x + 6$$
This equation is ready to be solved.

$3x + 2x - 4 = 5 + 2 - 3$
First simplify the left side of the equation. Combine $3x + 2x$.
$$3x + 2x = 5x$$
Substitute $5x$ for $3x + 2x$.
$$5x - 4 = 5 + 2 - 3$$
Next simplify the right side of the equation.
$$5 + 2 - 3 = 4$$
Substitute 4 for $5 + 2 - 3$.
$$5x - 4 = 4$$
This equation is ready to be solved.

Look how easy it is!

$4x + 2x - 7 + 9 + x = 5x - x$
First, simplify the left side of the equation. Combine all the terms with the same variable.
$$4x + 2x + x = 7x$$
Substitute $7x$ for $4x + 2x + x$.
$$7x - 7 + 9 = 5x - x$$

Combine the numbers on the left side of the equation.
$$-7 + 9 = 2$$
Substitute 2 for $-7 + 9$.
$$7x + 2 = 5x - x$$
Next simplify the right side of the equation. Subtract x from $5x$.
$$5x - x = 4x$$
Substitute $4x$ for $5x - x$.
$$7x + 2 = 4x$$
This equation is ready to be solved.

Caution—Major Mistake Territory!

In simplifying an equation so that it can be solved, whatever is on the left side of the equals sign stays on the left side.

Whatever is on the right side of the equals sign stays on the right side.

Don't mix the terms from the left side of the equation with the terms on the right side of the equation.

BRAIN TICKLERS
Set # 13

Simplify these equations. Remember to *Please Excuse My Dear Aunt Sally.*

1. $3(x + 2) = 0$

2. $5x + 1 + 2x = 4$

3. $6x = 4 - 1$

4. $3x - 2x = 6 - 7$

5. $2(x + 1) + 2x = 8$

6. $5x - 2x = 3(4 + 1)$

7. $x - 4x = 5 - 2 - 8$

8. $3x - 2x + x - 4 + 3 - 2 = 0$

(Answers are on page 95.)

Step 2: Using addition and subtraction to solve equations with one variable

Once an equation is simplified, the next step is to get all the variables on one side of the equation and all the numbers on the other side of the equation. To do this, you add and/or subtract the same number or variable from both sides of the equation.

If an equation is a true sentence, what is on one side of the equals sign is equal to what is on the other side of the equals sign. You can add the same number or variable to both sides of an equation—the new equation is still true. You can also subtract the same number from both sides of the equation to give a true equation.

Start by getting all the variables on the left side of the equation and all the numbers on the right side of the equation.

$x - 4 = 8$

All the variables are already on the left side of the equation. To get all the numbers on the right side of the equation, add four to both sides of the equation. Why 4? Because $-4 + 4 = 0$, and you will be left with x alone. When you add 4 to the left side of the equation, the -4 will disappear.

All the variables will be on the left side of the equation and all the numbers will be on the right side of the equation.

$$x - 4 + 4 = 8 + 4$$

Combine terms.

$$x = 12$$

$x + 5 = 12$

All the variables are already on the left side of the equation. To get all the numbers on the right side of the equation, subtract five from both sides of the equation. Why 5? Because $5 - 5 = 0$. When you subtract five from the left side of the equation, there will be no numbers on the left side of the equation.

$$x + 5 - 5 = 12 - 5$$

Combine terms.

$$x = 7$$

$2x = x - 7$

You want to get all the variables on the left side of the equation. To do this, subtract x from both sides of the equation. Why x? Because $x - x = 0$.

$$2x - x = x - 7 - x$$

Now combine like terms.

$$x = -7$$

$4x - 5 = 3x - 1$

There are variables on both sides of the equation. First get all the variables on the left side of the equation. Subtract $3x$ from both sides of the equation. Why $3x$? Because $3x - 3x = 0$.

$$4x - 5 - 3x = 3x - 3x - 1$$

Simplify the equation.

$$x - 5 = -1$$

Now all the variables are on the left side of the equation. There are numbers on both sides of the equation. Now get all the numbers on the right side of the equation. To do this, add five to both sides of the equation. Why 5? Because $5 - 5 = 0$.

$$x - 5 + 5 = -1 + 5$$

Combine terms.

$$x = 4$$

Caution—Major Mistake Territory!

Remember: Whatever you do to one side of an equation, you must do to the other side of the equation. You must treat both sides equally.

$$x - 5 = -5$$

If you add 5 to the left side of the equation, you must also add 5 to the right side of the equation. Don't just add 5 to one side.

$$x - 5 + 5 = -5 + 5$$
$$x = 0$$

BRAIN TICKLERS
Set # 14

Solve these equations by adding or subtracting the same number or variable from both sides. Keep the variable x on the left-hand side of the equation and the numbers on the right-hand side of the equation.

1. $x - 3 = 10$

2. $x - (-6) = 12$

3. $-5 + x = 4$

4. $x + 7 = 3$

5. $5 + 2x = 2 + x$

6. $3x + 3 = 3 + 2x$

7. $4x - 4 = -6 + 3x$

8. $5x - 2 = -8 + 4x$

(Answers are on page 95.)

Step 3: Using multiplication/division to solve equations

You can multiply one side of an equation by any number as long as you multiply the other side of the equation by the same number. Both sides of the equation will still be equal. You can also divide one side of an equation by any number as long as you divide the other side of the equation by the same number.

Once you have all the variables on one side of the equation and all the numbers on the other side of the equation, how do you decide what number to multiply or divide by? You pick the number that will give you only one x.

If the equation has $5x$, divide both sides of the equation by 5.
If the equation has $3x$, divide both sides of the equation by 3.
If the equation has $\frac{1}{2}x$, multiply both sides of the equation by $\frac{2}{1}$.
If the equation has $\frac{3}{5}x$, multiply both sides of the equation by $\frac{5}{3}$.

$4x = 8$

Divide both sides of the equation by four.

Why 4? Because $4 \div 4 = 1$, so you will only have one x.

$$4x \div 4 = 8 \div 4$$

Compute by dividing.

$$x = 2$$

$-2x = -10$

Divide both sides of the equation by (-2).

Why -2? Because $\frac{-2}{-2} = 1$.

$$\frac{-2x}{-2} = \frac{-10}{-2}$$

Compute by dividing.

$$x = 5$$

$\frac{1}{2}x = 10$

Multiply both sides of the equation by 2.

Why 2? Because $2\left(\frac{1}{2}\right) = 1$.

$$2\left(\frac{1}{2}x\right) = 2(10)$$

Compute by multiplying.

$$x = 20$$

$-\frac{5}{2}x = 10$

Multiply both sides of the equation by $-\frac{2}{5}$.

Why $-\frac{2}{5}$? Because $-\frac{2}{5}\left(-\frac{5}{2}\right) = 1$.

$$-\frac{2}{5}\left(\frac{5}{2}x\right) = -\frac{2}{5}(10)$$

Compute by multiplying.

$$x = -4$$

BRAIN TICKLERS
Set # 15

Solve these equations.

1. $\frac{1}{3}x = 2$

2. $3x = -3$

3. $4x = 0$

4. $\frac{1}{4}x = \frac{1}{2}$

5. $\frac{3}{2}x = -3$

(Answers are on page 96.)

SOLVING MORE EQUATIONS

SUM IT UP!

Remember the three steps for solving an equation.

Step 1: **Simplify each side of the equation.**
Use the Order of Operations.

Step 2: **Add and/or Subtract.**
Add and/or subtract the same number and/or variable from both sides of the equations.

Step 3: **Multiply or Divide.**
Multiply or divide both sides of the equation by the same number.

Here is an example of an equation to solve where you must use all three steps.

$3x - x + 3 = 7$

Step 1: Simplify the equation.
Combine like terms.
$2x + 3 = 7$

Step 2: Add and/or subtract.
Subtract 3 from both sides.
$2x + 3 - 3 = 7 - 3$
Compute by subtracting.
$2x = 4$

Step 3: Multiply and/or divide.
Divide both sides of the equation by 2.
$\frac{2x}{2} = \frac{4}{2}$
Compute by dividing.
$x = 2$

Here is another example. Remember the three steps.

Step 1: Simplify the equation.

Step 2: Add and/or subtract.

Step 3: Multiply and/or divide.

$3(x + 5) = 15 + 6$

Step 1: Simplify the equation.
Multiply $3(x + 5)$.
$3x + 15 = 15 + 6$
Combine like terms.
$3x + 15 = 21$

Step 2: Add and/or subtract.
Subtract 15 from both sides.
$3x + 15 - 15 = 21 - 15$
Compute by subtracting.
$3x = 6$

Step 3: Multiply and/or divide.
Divide both sides of the equation by 3.
$$\frac{3x}{3} = \frac{6}{3}$$
Compute by dividing.
$$x = 2$$

Here is another example.

$$2x - 5 + 3x - 1 = 3x + 4$$

Step 1: Simplify by combining like terms.
$$5x - 6 = 3x + 4$$

Step 2: Add 6 to both sides.
$$5x - 6 + 6 = 3x + 4 + 6$$
Compute by adding.
$$5x = 3x + 10$$

Subtract $3x$ from both sides.
$$5x - 3x = 3x + 10 - 3x$$
Compute by subtracting.
$$2x = 10$$

Step 3: Divide both sides by 2.
$$\frac{2x}{2} = \frac{10}{2}$$
Compute by dividing.
$$x = 5$$

Here is a fourth example.

$$x - \frac{1}{2}x + 5 - 3 = 0$$

Step 1: Simplify by combining like terms.
$$\frac{1}{2}x + 2 = 0$$

Step 2: Subtract 2 from both sides.
$$\frac{1}{2}x + 2 - 2 = 0 - 2$$
Compute. Do the subtraction.
$$\frac{1}{2}x = -2$$

Step 3: Multiply both sides by 2.

$$2\left(\tfrac{1}{2}x\right) = 2(-2)$$

Compute. Multiply $2(-2)$.

$$x = -4$$

BRAIN TICKLERS
Set # 16

Solve each of the following equations.

1. $3(x + 1) = 6$

2. $3x - 5x + x = 3 - 2x$

3. $5x + 3 + x = 3 - 6$

4. $\tfrac{1}{2}x + 5 = 6 - 2$

5. $\tfrac{2}{3}x + 1 = -5$

6. $5(2x - 2) = 3(x - 1) + 7$

(Answers are on page 97.)

Checking your work

Once you solve an equation, it is important to check your work. Substitute the answer into the original equation wherever you see an x. Compute the value of the sentence with the number in place of x. If both sides of the equation are equal, the answer is correct.

EXAMPLE: Bill solved the equation $3x + 1 = 10$. Bill came up with $x = 3$.

> To check, substitute 3 for x.
> $3(3) + 1 = 10$
>
> Compute. Multiply $3(3)$.
> $9 + 1 = 10$
>
> Compute. Add $9 + 1$.
> $10 = 10$
> Bill was right, $x = 3$.

EXAMPLE: Rosa solved $3(x + 5) = 10$. She came up with $x = 2$.

To check, substitute 2 for x.
$3(2 + 5) = 10$

Compute. Add $2 + 5$.
$3(7) = 10$

Compute. Multiply 3 times 7.
$21 = 10$
Because 21 does not equal ten, x cannot be 2.
Rosa made a mistake.

EXAMPLE: Jodie solved the equation $3x - 2x + 5 = 2$. She came up with $x = -3$.

To check, substitute -3 for x.
$3(-3) - 2(-3) + 5 = 2$

Multiply $3(-3)$ and $2(-3)$.
$-9 - (-6) + 5 = 2$

Change $- (-6)$ to $+6$.
$-9 + 6 + 5 = 2$

Add.
$2 = 2$
Jodie was right, $x = -3$.

EXAMPLE: Mike solved the equation $3(x + 2) = \frac{1}{2}(x - 2)$. Mike came up with $x = 4$.

To check, substitute 4 for x.

$3(4 + 2) = \frac{1}{2}(4 - 2)$

Add the numbers inside the parentheses.

$3(6) = \frac{1}{2}(2)$

Multiply $3(6)$ and $\frac{1}{2}(2)$.

$18 = 1$

Because 18 is not equal to 1, Mike was not correct; x is not equal to 4.

Caution—Major Mistake Territory!

When you check a problem, you find out whether your answer is right or wrong, but you do not find the correct answer if your answer is wrong. In order to find the correct answer, try to solve the problem again.

BRAIN TICKLERS
Set # 17

Three of these problems were solved incorrectly. Which ones were they? To find out, check each problem by substituting the answer for the variable.

1. $x + 7 = 10$ $x = 4$

2. $4x = 20$ $x = 5$

3. $2(x - 6) = 0$ $x = 0$

4. $3x + 5 = -4$ $x = -3$

5. $\frac{2}{3}x + 1 = -5$ $x = -9$

6. $4x - 2x - 7 = -1$ $x = -1$

(Answers are on page 98.)

WORD PROBLEMS

Solving word problems is simply a matter of knowing how to change Plain English into Math Talk! Once you write a problem down correctly, solving it is easy.

MATH TALK!

These simple rules should help you change Plain
English into Math Talk.

Rule 1: Change the word *equals* or any of the words *is, are,
was,* and *were* into an equals sign.

Rule 2: Use the letter x to represent the phrase "a number."

Let x = a number. Or, use the letter x for what you don't know.

EXAMPLE: A number is twice twelve.
　　　Change "a number" to x.
　　　Change "is" to "=."
　　　Change "twice twelve" to "2(12)."
　　　In Math Talk, the sentence is $x = 2(12)$.

EXAMPLE: John is two inches shorter than Quan. Quan is 62
inches tall. How tall is John?
　　　Change John to "x" for John's height.
　　　Change "is" to "=."
　　　Change "two inches shorter than Quan" to "Quan − 2."
　　　Change Quan to "62." John = Quan minus two inches.
　　　In Math Talk, the sentence is $x = 62 - 2$.

BRAIN TICKLERS
Set # 18

Change the following Plain English phrases into Math Talk.

1. five less than a number

2. three more than a number

3. four times a number

4. one fifth of a number

5. the difference between a number and three

6. the product of eight and a number

7. the sum of four and a number

(Answers are on page 99.)

Here is how we can begin to solve word problems.

PROBLEM 1: A number plus three is twelve. Find the number.

Change this sentence to Math Talk.
Change "a number" to "x."
Change "plus three" to "$+3$."
Change "is" to "$=$."
Change "twelve" to "12."
$x + 3 = 12$

Solve. Subtract three from both sides of the equation.
$x + 3 - 3 = 12 - 3$

Compute.
$x = 9$
The number is 9.

PROBLEM 2: Four times a number plus two is eighteen. Find the number.

Change this sentence to Math Talk.
Change "a number" to x.
Change "four times a number" to "$4x$."
Change "plus two" to "$+2$."
Change "is" to "$=$."
Change "eighteen" to "18."
$4x + 2 = 18$

Solve. Subtract two from both sides of the equation.
$4x + 2 - 2 = 18 - 2$

Compute. Do the subtraction.
$4x = 16$

Divide both sides of the equation by four.
$\frac{4x}{4} = \frac{16}{4}$
Compute. Divide.
$x = 4$
The number is 4.

PROBLEM 3: Two times the larger of two consecutive integers is three more than three times the smaller integer. Find both integers.

Change to Math Talk.
Clue: Two consecutive integers are x and $x + 1$; x is the smaller integer and $x + 1$ is the larger integer.

Change "two times the larger of two consecutive integers" to "$2(x + 1)$."
Change "is" to "$=$."
Change "three more" to $3+$.
Change "three times the smaller integer" to "$3x$."
$2(x + 1) = 3 + 3x$

Simplify by multiplication and the Distributive Property.
$2x + 2 = 3 + 3x$

Subtract $2x$ from both sides.
$2x - 2x + 2 = 3 + 3x - 2x$

Compute by subtracting.
$2 = 3 + x$

Subtract 3 from both sides.
$2 - 3 = 3 - 3 + x$

Compute by subtracting.
$x = -1$

The two consecutive integers are x and $x + 1$.
The two consecutive integers are -1 and 0.

SUPER BRAIN TICKLERS

Solve for x.

1. $4x - (2x - 3) = 0$

2. $5(x - 2) = 6(2x + 1)$

3. $4x - 2x + 1 = 5 + x - 7$

4. $\frac{1}{2}x = \frac{1}{4}x + 2$

5. $6(x - 2) - 3(x + 1) = 4(3 + 2)$

(Answers are on page 99.)

BRAIN TICKLERS— THE ANSWERS

Set # 13, page 77

1. $3x + 6 = 0$

2. $7x + 1 = 4$

3. $6x = 3$

4. $x = -1$

5. $4x + 2 = 8$

6. $3x = 15$

7. $-3x = -5$

8. $2x - 3 = 0$

Set # 14, page 80

1. $$x - 3 = 10$$
$$x - 3 + 3 = 10 + 3$$
$$x = 13$$

2. $$x - (-6) = 12$$
$$x - (-6) + (-6) = 12 - 6$$
$$x = 6$$

3. $$-5 + x = 4$$
$$-5 + 5 + x = 4 + 5$$
$$x = 9$$

4. $$x + 7 = 3$$
$$x + 7 - 7 = 3 - 7$$
$$x = -4$$

5.
$$5 + 2x = 2 + x$$
$$5 - 5 + 2x - x = 2 - 5 + x - x$$
$$x = -3$$

6.
$$3x + 3 = 3 + 2x$$
$$3x - 2x + 3 - 3 = 3 - 3 + 2x - 2x$$
$$x = 0$$

7.
$$4x - 4 = -6 + 3x$$
$$4x - 3x - 4 + 4 = -6 + 4 + 3x - 3x$$
$$x = -2$$

8.
$$5x - 2 = -8 + 4x$$
$$5x - 4x - 2 + 2 = -8 + 2 + 4x - 4x$$
$$x = -6$$

Set # 15, page 83

1. $\frac{1}{3}x = 2$

$$3\left(\frac{1}{3}x\right) = 3(2)$$
$$x = 6$$

2. $3x = -3$

$$3\left(\frac{x}{3}\right) = \frac{-3}{3}$$
$$x = -1$$

3. $4x = 0$

$$\frac{4x}{4} = \frac{0}{4}$$
$$x = 0$$

4. $\frac{1}{4}x = \frac{1}{2}$

$$4\left(\frac{1}{4}x\right) = 4\left(\frac{1}{2}\right)$$

$$x = 2$$

5. $\frac{3}{2}x = -3$

$$\frac{2}{3}\left(\frac{3}{2}x\right) = \frac{2}{3}(-3)$$

$$x = -2$$

Set # 16, page 86

1. $3(x + 1) = 6$

$$3x + 3 = 6$$
$$3x = 3$$
$$x = 1$$

2. $3x - 5x + x = 3 - 2x$

$$-x = 3 - 2x$$
$$x = 3$$

3. $5x + 3 + x = 3 - 6$

$$6x + 3 = -3$$
$$6x = -6$$
$$x = -1$$

4. $\frac{1}{2}x + 5 = 6 - 2$

$$\frac{1}{2}x + 5 = 4$$

$$\frac{1}{2}x = -1$$

$$x = -2$$

5. $\frac{2}{3}x + 1 = -5$

$$\frac{2}{3}x = -6$$
$$x = -9$$

6. $5(2x - 2) = 3(x - 1) + 7$

$10x - 10 = 3x - 3 + 7$

$10x - 10 = 3x + 4$

$$7x = 14$$
$$x = 2$$

Set # 17, page 90

1. $x + 7 = 10; x = 4$

$(4) + 7 = 10$

$10 = 11$

This problem is solved incorrectly.

2. $4x = 20; x = 5$

$4(5) = 20$

$20 = 20$

Correct.

3. $2(x - 6) = 0; x = 0$

$2(0 - 6) = 0$

$2(-6) = 0$

$-12 = 0$

This problem is solved incorrectly.

4. $3x + 5 = -4; x = -3$

$3(-3) + 5 = -4$

$-9 + 5 = -4$

$-4 = -4$

Correct.

5. $\frac{2}{3}x + 1 = -5; x = -9$

$\frac{2}{3}(-9) + 1 = -5$

$-6 + 1 = -5$

Correct.

6. $4x - 2x - 7 = -1; x = -1$

$4(-1) - 2(-1) - 7 = -1$

$-4 + 2 - 7 = -1$

$-9 = -1$

This problem is solved incorrectly.

Set # 18, page 92

1. $x - 5$

2. $x + 3$

3. $4x$

4. $\frac{1}{5}x$ or $\frac{x}{5}$

6. $x - 3$

7. $4 + x$ or $x + 4$

Super Brain Ticklers, page 94

1. $x = -\frac{3}{2}$

2. $x = -\frac{16}{7}$

3. $x = -3$

4. $x = 8$

5. $x = \frac{35}{3}$

CHAPTER FOUR

Solving Inequalities

An inequality is a sentence in which one side of the expression is greater than or less than the other side of the expression. Inequalities have symbols for the four different expressions.

> means greater than.

< means less than.

≥ means greater than or equal to.

≤ means less than or equal to.

Notice that the symbol for greater than or equal to, ≥, is just the symbol for greater than with half an equals sign on the bottom. The symbol for less than or equal to, ≤ is just the symbol for less than with half an equals sign on the bottom. Notice that the symbol ≥ is read "greater than *or* equal to," not "greater than *and* equal to." No number can be greater than and equal to another number at the same time.

MATH TALK!

Change the following inequalities from Math Talk into Plain English.

$4 > 0$
Four is greater than 0.

$-3 \geq -7$
Negative three is greater than or equal to negative seven.

$2 < 5$
Two is less than five.

$-6 \leq 4$
Negative six is less than or equal to four.

$8 \leq 8$
Eight is less than or equal to eight.

An inequality can be true or false. Here are some examples of true inequalities.

The inequality $3 > 1$ is true, since three is greater than one.
The inequality $5 < 10$ is true, since five is less than ten.
The inequality $-6 \leq -6$ is true if negative six is equal to or less than negative six. Since negative six is equal to negative six, this sentence is true.

Here are some examples of false inequalities.

The inequality $5 > 10$ is not true, since five is not greater than ten.

The inequality $-6 \leq -9$ is not true, since negative six is not less than negative nine and negative six is not equal to negative nine.

BRAIN TICKLERS
Set # 19

Are the following statements true or false?

_____ 1. $3 > 4$ _____ 4. $-6 \leq -1$

_____ 2. $6 \geq 6$ _____ 5. $0 \leq 0$

_____ 3. $0 < -4$

(Answers are on page 126.)

Sometimes one side of an inequality has a variable. The inequality states whether the variable is larger, smaller, or maybe even equal to a specific number.

MATH TALK!

Here is how you change the following inequalities from Math Talk into Plain English.

$$x > 4$$
A mystery number is greater than four

$$x < -1$$
A mystery number is less than negative one.

$$x \leq -2$$
A mystery number is less than or equal to negative two.

$$x \geq 0$$
A mystery number is greater than or equal to zero.

If an inequality has a variable in it, some numbers will make it true while others will make it false. For example, consider $x > 2$. If x is equal to 3, 4, or 5, this inequality is true. It is even true if x is equal to $2\frac{1}{2}$. But $x > 2$ is false if x is equal to 0, -1, or -2. The inequality $x > 2$ is even false if x is equal to two, since two cannot be greater than two.

When an inequality says, for example, "x is greater than one $(x > 1)$," it means that x can be any number greater than one, not just any whole number greater than one. If $x > 1$, then x can be 1.1 or 1.234 or 1,000,001.5, or x could even be the square root of two.

GRAPHING INEQUALITIES

Often, inequalities with a variable are graphed. The graph gives you a quick picture of all the mystery numbers on the number line that will work. There are three steps to graphing an inequality on the number line.

Step 1: Locate the number in the inequality on the number line.

Step 2: If the inequality is either $>$ or $<$, circle the number. If the inequality is either $\geq$ or $\leq$, circle and shade the number.

Step 3: If the inequality is either $>$ or $\geq$, shade the number line to the right of the number. If the inequality is either $<$ or $\leq$, shade the number line to the left of the number.

Now let's try an example.

$x > 2$

Step 1: Locate the number in the inequality on the number line. Two is marked on the number line.

Step 2: If the inequality is either $>$ or $<$, circle the number. If the inequality is either $\geq$ or $\leq$, circle and shade the number.
Circle the number two. When the number is circled, it means it is not included in the graph.

Step 3: If the inequality is either $>$ or $\geq$, shade the number line to the right of the number.
If the inequality is either $<$ or $\leq$, shade the number line to the left of the number.
Since the inequality is $x > 2$, shade the number line to the right of the number two. Notice that all the numbers are shaded, not just all the whole numbers. All the numbers to the right of two are greater than two.

Here is another example:

$x \leq -1$

Step 1: Locate the number in the inequality on the number line. Negative one is marked on the number line.

Step 2: If the inequality is either $>$ or $<$, circle the number. If the inequality is either $\geq$ or $\leq$, circle and shade the number.
Since the inequality is $x \leq -1$, circle and shade -1. Circling and shading -1 means -1 is included in the graph.

Step 3: If the inequality is either $>$ or $\geq$, shade the number line to the right of the number.
If the inequality is either $<$ or $\leq$, shade the number line to the left of the number.
Since the graph is $x \leq -1$, shade to the left of -1.

Here is a third example:

$y \geq 0$

Step 1: Locate the number in the inequality on the number line. Zero is already marked on the number line.

Step 2: If the inequality is either $>$ or $<$, circle the number. If the inequality is either $\geq$ or $\leq$, circle and shade the number. Circle and shade zero, since it is included in the graph.

Step 3: If the inequality is either $>$ or $\geq$, shade the number line to the right of the number. If the inequality is either $<$ or $\leq$, shade the number line to the left of the number. Since the equation is $y \geq 0$, the entire number line to the right of 0 is shaded.

MATH TALK!

Here is how you read these graphs and change them from Math Talk to Plain English.

x is greater than or equal to negative two.
$$x \geq -2$$

x is greater than three.
$$x > 3$$

x is less than or equal to one half.
$$x \leq \frac{1}{2}$$

x is less than negative three halves.
$$x < -\frac{3}{2}$$

BRAIN TICKLERS
Set # 20

Graph the following inequalities.

_____ 1. $x > 2$

_____ 2. $x < -2$

_____ 3. $x \geq 2$

_____ 4. $x \geq -2$

_____ 5. $x \leq -2$

_____ 6. $x < 2$

(Answers are on page 126.)

SOLVING INEQUALITIES

Solving inequalities is *almost* exactly like solving equations. There are a couple of important differences, so pay close attention.

Follow the same three steps when solving inequalities that you followed when solving equations, and add a fourth step.

Step 1: Simplify each side of the inequality using the Order of Operations. Simplifying each side of the inequality is a two-step process. First simplify the left side. Next simplify the right side.

Step 2: Add and/or subtract numbers and/or variable terms from both sides of the inequality. Move all the variables to one side of the inequality and all the numbers to the other side of the inequality.

Step 3: Multiply or divide both sides of the inequality by the same number. Now here is the key difference: *If you multiply or divide by a negative number, reverse the direction of the inequality.*

Step 4: Graph the answer on the number line.

Now you can solve an inequality. Here are three examples.

$2(x - 1) > 4$

Step 1: Simplify each side of the inequality.
Multiply $(x - 1)$ by 2.
$2(x - 1) = 2x - 2$
The inequality is now $2x - 2 > 4$.

Step 2: Add or subtract the same number from both sides of the inequality.
Add 2 to both sides of the inequality. Why 2? Now all the numbers are on the right side of the inequality.
$2x - 2 + 2 > 4 + 2$

Simplify.
$2x > 6$

Step 3: Multiply or divide both sides of the inequality by the same number.
Divide both sides of the equation by 2. Why 2? Because you want to have one x on one side of the equation.
$\frac{2x}{2} > \frac{6}{2}$
$x > 3$

Step 4: Graph the answer on the number line.
Circle the number 3 on the number line. Circling 3 indicates 3 is not included in the graph.
Shade the number line to the right of the number 3, since the inequality reads x is greater than 3.
All the numbers greater than 3 will make the inequality $2(x - 1) > 4$ true.

$$2x - 5x + 4 \leq 10$$

Step 1: Simplify each side of the inequality.
Combine the like terms on the left side of the inequality.
$$-3x + 4 \leq 10$$

Step 2: Add or subtract the same number from both sides of the inequality.
Subtract 4 from both sides of the inequality. Why 4? So that all the numbers are on the right side of the inequality.
$$-3x + 4 - 4 \leq 10 - 4$$
Compute.
$$-3x \leq 6$$

Step 3: Multiply or divide both sides of the equation by the same number.
Divide both sides of the inequality by -3. Why -3? Because you want to have only one x on the left side of the equation. Because you are dividing by a negative number, you must reverse the direction of the inequality.
$$\frac{-3x}{-3} \geq \frac{6}{-3}$$
Compute.
$$x \geq -2$$

Step 4: Graph the answer on the number line.
Circle the number -2 on the number line. Shade the circle. Circling and shading -2 says it is included in the graph.
Shade the number line to the right of the number -2, since the inequality reads x is greater than -2.
All the numbers greater than or equal to -2 will make the inequality $2x - 5x + 4 \leq 10$ true.

Caution—Major Mistake Territory!

Whenever you multiply or divide an inequality by a negative number, you must remember to change the direction of that inequality. When you multiply or divide an inequality by a negative number,

$>$ is changed to $<$;

$\geq$ is changed to $\leq$;

$<$ is changed to $>$;

$\leq$ is changed to $\geq$.

Example: $-4x > 16$

To solve, divide both sides by -4.

$$\frac{-4x}{-4} > \frac{16}{-4}$$
$$x < -4$$

$-\frac{1}{2}(x - 10) \geq 7$

Step 1: Simplify each side of the inequality.

Simplify the left side of the inequality. Multiply $(x - 10)$ by $-\frac{1}{2}$. Since you are only multiplying one side of the equation by a negative number $\left(-\frac{1}{2}\right)$, don't reverse the inequality sign.

$-\frac{1}{2}x + 5 \geq 7$

There is nothing to simplify on the right side of the inequality.

Step 2: Add or subtract the same number from both sides of the inequality.

Subtract 5 from both sides of the inequality. Why 5? Because you want all the numbers on the right side.

$-\frac{1}{2}x + 5 - 5 \geq 7 - 5$

Compute.

$-\frac{1}{2}x \geq 2$

Step 3: Multiply or divide both sides of the inequality by the same number.

Multiply both sides of the inequality by -2. Why -2? Because you want to have only one x on the left side of the inequality. Remember to reverse the direction of the inequality, since you are multiplying both sides of the inequality by a negative number.

$$-2\left(-\tfrac{1}{2}x\right) \le -2(2)$$

Compute.

$x \le -4$

Step 4: Graph the answer on the number line.

Circle -4 on the number line. Shade in the circle. Shading says -4 is included in the graph.

Shade the number line to the left of -4, since the equation reads x is less than -4.

All the numbers less than or equal to -4 make the equation $-\tfrac{1}{2}(x - 10) \ge 7$ true.

BRAIN TICKLERS
Set # 21

All of these inequalities require that you multiply or divide both sides of the equation by the same number to find the solution. Some require a sign change and some don't. Put a "C" next to the inequalities where you must change the sign.

_____ 1. $3x > 9$

_____ 2. $-2x \geq 10$

_____ 3. $-\frac{3}{2}x < -12$

_____ 4. $\frac{1}{2}x \leq 4$

_____ 5. $5x > -15$

_____ 6. $-2x > 0$

_____ 7. $\frac{1}{3}x \geq 9$

(Answers are on page 127.)

SUM IT UP!

When solving an inequality with one variable, follow these steps to success.

Step 1: Simplify both sides of the inequality. Combine like terms.

Step 2: Add or subtract the same number or variable on both sides of the inequality. Make sure all of the variables are on one side of the inequality and all the numbers are on the other side of the inequality.

Step 3: Multiply or divide both sides of the inequality by the same number. The objective is to get x on one side of the inequality by itself. If you multiply or divide by a negative number, you must reverse the direction of the inequality.

BRAIN TICKLERS
Set # 22

Solve these inequalities.

1. $3x + 5 > 7$

2. $-\frac{1}{2}x - 2 > 8$

3. $4(x - 2) < 8$

4. $-5(x - 1) < 5(x + 1)$

5. $4x + 2 - 4x \geq 5 - x - 4$

6. $\frac{1}{3}x - 2 \leq \frac{2}{3}x - 6$

(Answers are on page 127.)

Checking your work

To check to see if you solved an inequality correctly, follow these simple steps.

Step 1: Change the inequality sign in the problem to an equals sign.

Step 2: Substitute the number in the answer for the variable. If the sentence is true, continue to the next step. If the sentence is not true, STOP. The answer is wrong.

Step 3: Substitute zero into the inequality to check the direction of the inequality.

> **EXAMPLE:** The problem was $x + 3 > 4$. Allison thinks the answer is $x > 1$. Check to see if Allison is right or wrong.
>
> > First change the problem to an equation.
> > $x + 3 > 4$ becomes $x + 3 = 4$.
> > Substitute the answer for x.
> > Substitute 1 for x.
> > $1 + 3 = 4$
> > This is a true sentence.
> >
> > Now check the direction of the inequality by substituting 0 for x.
> > $0 + 3 > 4$
> > This is not true, so 0 is not part of the solution. If zero were part of the solution set, the answer would be $x < 1$. But because zero is not part of the solution, the correct answer is $x > 1$.
> > Allison was correct.

WORD PROBLEMS

The trickiest part of solving word problems with inequalities is changing the problem from Plain English to Math Talk. Much of the problem is changed the same way you change a word problem into an equation. The trick is deciding which inequality to use and which way it should point. Here are some tips that should help.

Look for these phrases when solving inequalities.

Whenever the following phrases are used, insert ">."

. . . is more than . . .
. . . is bigger than . . .
. . . is greater than . . .
. . . is larger than . . .

Whenever the following phrases are used, insert "<."

. . . is less than . . .
. . . is smaller than . . .

Whenever the following phrases are used, insert "≥."

. . . greater than or equal to . . .
. . . at least . . .

Whenever the following phrases are used, insert "≤."

. . . less than or equal to . . .
. . . at most . . .

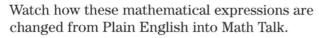

MATH TALK!

Watch how these mathematical expressions are changed from Plain English into Math Talk.

Twice a number is at most six.
$$2x \leq 6$$

Two plus a number is at least four.
$$2 + x \geq 4$$

Three less than a number is less than three times the number.
$$x - 3 < 3x$$

One fourth of a number is greater than three.
$$\tfrac{1}{4}x > 3$$

Here are four word problems that are solved. Study each of them.

PROBLEM 1: The product of five and a number is less than zero.

First change the problem from Plain English to Math Talk!
Change "the product of five and a number" to "$5x$."
Change "less than" to "$<$."
Change "zero" to "0."
In Math Talk, the problem now reads $5x < 0$.

Divide both sides by 5.
$$\frac{5x}{5} < \frac{0}{5}$$
$$x < 0$$
The mystery number is less than zero.

PROBLEM 2: The sum of two consecutive numbers is at least thirteen. What numbers could the first number be?

First change the problem from Plain English to Math Talk.
Change "the sum of two consecutive numbers" to "$x + (x + 1)$."
Change "is at least" to "$\geq$."
Change "thirteen" to "13."
In Math Talk, the problem now reads $x + (x + 1) \geq 13$.

Solve the problem. First simplify.
$x + x + 1 \geq 13$ becomes $2x + 1 \geq 13$.
Subtract one from both sides of the equation.
$$2x + 1 - 1 \geq 13 - 1$$
$$2x \geq 12$$
$$x \geq 6$$

The first of the two consecutive numbers must be greater than or equal to six.

PROBLEM 3: Three times a number plus one is at most ten. What could the number be?

First change the problem from Plain English to Math Talk.
Change "three times a number" to "$3x$."
Change "plus one" to "$+1$."
Change "is at most" to "$\leq$."
Change "ten" to "10."
In Math Talk, the problem now reads $3x + 1 \leq 10$.

Now solve the inequality.
Subtract one from both sides of the inequality.
$$3x + 1 - 1 \leq 10 - 1$$
$$3x \leq 9$$

Divide both sides by three.
$$\frac{3x}{3} \leq \frac{9}{3}$$
$$x \leq 3$$
The mystery number is at most three.

PROBLEM 4: One half of a number is greater than five. What is the number?

First change the problem from Plain English to Math Talk.
Change "one half of a number" to "$\frac{1}{2}x$."
Change "is greater than" to ">."
Change "five" to "5."
In Math Talk, the problem reads $\frac{1}{2}x > 5$.

Now solve the inequality.
Multiply both sides of the inequality by two.
$$2\left(\frac{1}{2}x\right) > 2(5)$$
$$x > 10$$
The mystery number is greater than ten.

SUPER BRAIN TICKLERS

Solve for x.

1. $5(x - 2) > 6(x - 1)$

2. $3(x + 4) < 2x - 1$

3. $\frac{1}{4}x - \frac{1}{2} < \frac{1}{2}$

4. $-2(x - 3) > 0$

5. $-\frac{1}{2}(2x + 2) < 5$

(Answers are on page 129.)

BRAIN TICKLERS—
THE ANSWERS

Set # 19, page 105

1. False

2. True

3. False

4. True

5. True

Set # 20, page 112

Graph the following inequalities.

_____ 1. $x > 2$

_____ 2. $x < -2$

_____ 3. $x \geq 2$

_____ 4. $x \geq -2$

_____ 5. $x \leq -2$

_____ 6. $x < 2$

Set # 21, page 118

1. (no change)

2. C

3. C

4. (no change)

5. (no change)

6. C

7. (no change)

Set # 22, page 120

1.
$$3x + 5 > 7$$
$$3x + 5 - 5 > 7 - 5$$
$$3x > 2$$
$$\frac{3x}{3} > \frac{2}{3}$$
$$x > \frac{2}{3}$$

2.
$$-\tfrac{1}{2}x - 2 > 8$$
$$-\tfrac{1}{2}x - 2 + 2 > 8 + 2$$
$$-\tfrac{1}{2}x > 10$$
$$(-2)\left(-\tfrac{1}{2}x\right) < (-2)(10)$$
$$x < -20$$

3. $4(x - 2) < 8$

$4x - 8 < 8$

$4x - 8 + 8 < 8 + 8$

$4x < 16$

$x < 4$

4. $-5(x - 1) < 5(x + 1)$

$-5x + 5 < 5x + 5$

$-5x + 5 - 5 < 5x + 5 - 5$

$-5x < 5x$

$-5x + 5x < 5x + 5x$

$0 < 10x$

$\frac{0}{10} < \frac{10x}{10}$

$0 < x$

5. $4x + 2 - 4x \geq 5 - x - 4$

$2 \geq 1 - x$

$2 - 1 \geq 1 - 1 - x$

$1 \geq -x$

$-1(1) \leq -1(-x)$

$-1 \leq x$

6. $\frac{1}{3}x - 2 \leq \frac{2}{3}x - 6$

$\frac{1}{3}x - 2 + 2 \leq \frac{2}{3}x - 6 + 2$

$\frac{1}{3}x \leq \frac{2}{3}x - 4$

$\frac{1}{3}x - \frac{2}{3}x \leq \frac{2}{3}x - \frac{2}{3}x - 4$

$-\frac{1}{3}x \leq -4$

$(-3)\left(-\frac{1}{3}x\right) \geq (-3)(-4)$

$x \geq 12$

Super Brain Ticklers, page 125

1. $x < -4$

2. $x < -13$

3. $x < 4$

4. $x < 3$

5. $x > -6$

CHAPTER FIVE

Exponents

Exponents are a shorthand way of writing repeated multiplication. The expression $2 \cdot 2 \cdot 2 \cdot 2 \cdot 2 \cdot 2$ is read two times two times two times two times two times two. But instead of writing and reading all that, mathematicians just say 2^6, which is two to the sixth power. The expression 2^6 (called an *exponential expression*) means to multiply 2 by itself six times. In the expression 2^6, 2 is the *base* and 6 is the *exponent*. Think of exponents as shorthand. They are a short way of writing repeated multiplication.

MATH TALK!

Watch how to change these expressions from Math Talk to Plain English.

$$4^2$$
four to the second power
four squared
four times four

$$5^3$$
five to the third power
five cubed
five times five times five

$$6^4$$
six to the fourth power
six times six times six times six

$$x^5$$
x to the fifth power
x times x times x times x times x

BRAIN TICKLERS
Set # 23

Match the following multiplication problems to their exponential form.

_____ 1. (3)(3)(3)(3)   A. x^4

_____ 2. $7 \cdot 7 \cdot 7 \cdot 7 \cdot 7$ B. 5^3

_____ 3. two squared   C. 7^5

_____ 4. 4(4)   D. 3^4

_____ 5. $(x)(x)(x)(x)$ E. 2^2

_____ 6. five cubed   F. 4^2

(Answers are on page 158.)

Caution—Major Mistake Territory!

3^2 is three times three, which is nine.
$$3^2 = 3(3) = 9$$

3^2 is not three times two, which is six.
$$3^2 \neq 3(2) = 6$$

You can compute the value of exponential expressions by multiplying.

$$2^2 = 2(2) = 4$$
$$2^3 = 2(2)(2) = 8$$
$$2^4 = 2(2)(2)(2) = 16$$
$$2^5 = 2(2)(2)(2)(2) = 32$$
$$3^2 = 3 \times 3 = 9$$
$$3^3 = 3 \times 3 \times 3 = 27$$

When you square a negative number, the answer is always positive. A negative number times a negative number is a positive number.

$$(-2)^2 = (-2)(-2) = +4$$
$$(-3)^2 = (-3)(-3) = +9$$

A negative number cubed is always a negative number.

$$(-2)^3 = (-2)(-2)(-2) = -8$$
$$(-3)^3 = (-3)(-3)(-3) = -27$$

A negative number raised to any even power is always a positive number. A negative number raised to any odd power is always a negative number.

$$(-1)^{17} = -1$$
$$(-1)^{10} = +1$$
$$(-1)^{25} = -1$$
$$(-1)^{99} = -1$$
$$(-1)^{100} = 1$$
$$(-2)^2 = 4$$
$$(-2)^3 = -8$$
$$(-2)^4 = 16$$
$$(-2)^5 = -32$$
$$(-2)^6 = 64$$
$$(-2)^7 = -128$$

Zero power

Any number to the zero power is one.

> **EXAMPLE:**
> $$6^0 = 1$$
> $$4^0 = 1$$
> $$a^0 = 1$$

BRAIN TICKLERS
Set # 24

Compute the value of these exponential expressions.

1. 5^2

2. 2^6

3. 10^2

4. 4^3

5. 5^0

6. $(-3)^2$

7. $(-5)^3$

8. $(-4)^2$

9. $(-1)^5$

10. $(-1)^{12}$

(Answers are on page 158.)

MULTIPLYING EXPONENTS WITH COEFFICIENTS

Some exponential expressions have coefficients in front of them.
The exponential expression is multiplied by its coefficient.
In the expression $3(5)^2$,

    3 is the coefficient;
    5 is the base;
    2 is the exponent.

In the expression $2y^3$,

2 is the coefficient;
y is the base;
3 is the exponent.

In the expression $(5x)^2$,

1 is the coefficient;
$5x$ is the base;
2 is the exponent.

To compute the value of an exponential expression with a coefficient, compute the exponential expression first. Next, multiply by the coefficient. Compute the value of the following exponential expressions.

$5(3)^2$
First square the three.
$$3^2 = 9$$
Substitute 9 for 3^2.
$$5(9) = 45$$
So, $5(3)^2 = 45$.

$-3(-2)^2$
First square negative two.
$$(-2)(-2) = 4$$
Substitute 4 for $(-2)^2$.
$$-3(4) = -12$$
So, $-3(-2)^2 = -12$.

$2(5 - 3)^2$
Do what is inside the parentheses first.
$$5 - 3 = 2$$
Substitute 2 for $5 - 3$.
$$2(2)^2$$
Square the two.
$$2^2 = 4$$
Substitute 4 for 2^2.
$$2(4)$$
Multiply.
$$2(4) = 8$$
So, $2(5 - 3)^2 = 8$.

BRAIN TICKLERS
Set # 25

Compute the value of the following exponential expressions.

1. $3(5)^2$

2. $-4(3)^2$

3. $2(-1)^2$

4. $3(-1)^3$

5. $5(-2)^2$

6. $-\frac{1}{2}(-4)^2$

7. $-2(-3)^2$

8. $-3(-3)^3$

(Answers are on page 159.)

Caution—Major Mistake Territory!

The order in which you do mathematical operations involving exponents makes a difference in the answer.

For example $3(2)^2$ is not the same as $(3 \cdot 2)^2$.

The first, $3(2)^2$, is equal to $3(4)$, which is 12.
The second, $(3 \cdot 2)^2$, is equal to 6^2, which is 36.

ADDING AND SUBTRACTING EXPONENTIAL EXPRESSIONS

You can add and subtract exponential expressions if they have the same base and the same exponent. Just add or subtract the coefficients.

$2(a^3) + 5(a^3)$
Check to make sure the expressions have the same base and the same exponent.

> The letter a is the base for both.
> The number 3 is the exponent for both.

Next add the coefficients: $2 + 5 = 7$
So, $2(a^3) + 5(a^3) = 7(a^3)$.

Add $3(5^2)$ and $2(5^2)$.
First check to make sure the expressions have the same base and the same exponent.

> The number 5 is the base for both.
> The number 2 is the exponent for both.

Next add the coefficients: $3 + 2 = 5$.
So, $3(5^2) + 2(5^2) = 5(5^2)$.

$4(y^2) - 2(y^2)$

Check to make sure the equations have the same base and the same exponent.

The variable y is the base for both.
The number 2 is the exponent for both.

Next, subtract the coefficients: $4 - 2 = 2$.
So, $4(y^2) - 2(y^2) = 2y^2$.

$7(2^8) - 7(2^8)$

Check to make sure the expressions have the same base and the same exponent.

The number 2 is the base for both.
The number 8 is the exponent for both.

Next subtract the coefficients: $7 - 7 = 0$.
So, $7(2^8) - 7(2^8) = 0(2^8)$.
Since zero times any number is zero, $0(2^8) = 0$.

Caution—Major Mistake Territory!

You can only add and subtract exponential expressions if their bases and exponents are exactly the same.

What is $5(2^4) + 5(2^3)$?

You cannot add these two expressions. They have the same base, but they do not have the same exponent.

What is $4(x^4) - 3(y^4)$?

You cannot subtract these two expressions. They have the same exponent, but they do not have the same base.

BRAIN TICKLERS
Set # 26

Simplify each of the following problems by adding or subtracting.

1. $3(3)^2 + 5(3)^2$

2. $4(16)^3 - 2(16)^3$

3. $3x^2 - 5x^2$

4. $2x^0 + 5x^0$

5. $5x^4 - 5x^4$

(Answers are on page 159.)

MULTIPLYING EXPONENTIAL EXPRESSIONS

You can multiply two exponential expressions if they have the same base.

Just add the exponents.

Simplify $3^3 \cdot 3^2$.
Three is the base of both 3^3 and 3^2.
Just add the exponents.
So, $3^3 \cdot 3^2 = 3^{3+2} = 3^5 = 243$.

Simplify $(4)^5(4)^{-3}$.
$(4)^5$ and $(4)^{-3}$ both have the same base.
To simplify this expression, just add the exponents.
So, $(4)^5(4)^{-3} = 4^{5-3} = 4^2 = 16$.

Simplify $(5)^{10}(5)^{-10}$.
$(5)^{-10}$ and $(5)^{10}$ both have the same base.
To simplify this expression, just add the exponents.
So, $(5)^{-10}(5)^{10} = 5^{-10+10} = 5^0 = 1$.

Simplify a^3a^4.
a^3 and a^4 both have the same base.
To simplify, just add the exponents.
So, $a^3a^4 = a^{3+4} = a^7$.

You can multiply several terms together. Just add the exponents of all the terms that have the same base.

Simplify $6^3 \cdot 6^5 \cdot 6^{-2} \cdot 6^4$.
All these terms have the same base.
To simplify this expression, just add all the exponents.
So, $6^3 \cdot 6^5 \cdot 6^{-2} \cdot 6^4 = 6^{(3+5-2+4)} = 6^{10}$.

You can even multiply together exponential expressions with coefficients, as long as they have the same base. For example, you can multiply $4x^3$ times $6x^5$. Just follow these three *painless* steps.

Step 1: Multiply the coefficients together.

Step 2: Add the exponents.

Step 3: Combine the terms. Put the new coefficient first, the base second, and the new exponent third.

Simplify $3x^24x^5$.
First, multiply the coefficients together: $3(4) = 12$.
Second, add the exponents: $2 + 5 = 7$.
Now combine the terms. Put the new coefficient first, the base second, and the new exponent third.
So, $3x^24x^5 = 12x^7$.

Simplify $-6x(3x^3)$.
First, multiply the coefficients together. $(-6)(3) = -18$.
Second, add the exponents. $1 + 3 = 4$
Next combine the new terms. Put the new coefficient first, the base second, and the new exponent third.
So, $-6x(3x^3) = -18x^4$.

Caution—Major Mistake Territory!

You cannot multiply exponential expressions with different bases.

$3^2 \cdot 2^3$ cannot be simplified, because 3^2 and 2^3 do not have the same base.

$a^2 b^5$ cannot be simplified, because a^2 and b^5 do not have the same base.

BRAIN TICKLERS
Set # 27

Simplify the following exponential expressions.

1. $2^3 2^3$

2. $2^5 2^2$

3. $2^{10} 2^{-2}$

4. $2^{-1} \cdot 2^3 \cdot 2^1$

5. $x^3 x^{-2}$

6. $x^4 \cdot x^{-4}$

7. $6x^4(x^{-2})$

8. $-7x^2(5x^3)$

9. $(-6x^3)(-2x^{-3})$

(Answers are on page 159.)

DIVIDING EXPONENTIAL EXPRESSIONS

You can divide exponential expressions with the same base. Just subtract the exponents.

Simplify $3^3 \div 3^2$.
3^3 and 3^2 have the same base.
To simplify, subtract the exponents.
So, $3^3 \div 3^2 = 3^{\,3-2} = 3^1$.

Simplify $\dfrac{5^2}{5^3}$.

5^2 and 5^3 have the same base.
To simplify, subtract the exponents.
So, $\dfrac{5^2}{5^3} = 5^{2-3} = 5^{-1}$.

Simplify $\dfrac{x^4}{x^{-4}}$.

x^4 and x^{-4} have the same base.
To simplify, subtract the exponents.
So, $\dfrac{x^4}{x^{-4}} = x^{\,4-(-4)} = x^8$.

Simplify $\dfrac{a^{-3}}{a^2}$

a^{-3} and a^2 have the same base.
To simplify, subtract the exponents.
So, $\dfrac{a^{-3}}{a^2} = a^{-3-2} = a^{-5}$.

You can even divide exponential expressions with coefficients, as long as they have the same base. For example, you can divide $2x^3$ by $5x^5$. Just follow these three *painless* steps.

Only three
painless steps!

Step 1: Divide the coefficients.

Step 2: Subtract the exponents.

Step 3: Combine the terms. Put the new coefficient first, the base second, and the new exponent third.

Simplify $6x^4 \div 3x^2$.
First divide the coefficients: $6 \div 3 = 2$
Next subtract the exponents: $4 - 2 = 2$.
Combine the terms. Put the new coefficient first, the base second, and the new exponent third.
So, $6x^4 \div 3x^2 = 2x^2$.

Simplify $\dfrac{4x^3}{16x^{-2}}$.

First divide the coefficients 4 and 16: $\dfrac{4}{16} = \dfrac{1}{4}$.

Next subtract the exponents. $3 - (-2) = 5$.

Combine the terms. Put the new coefficient first, the base second, and the new exponent third.

So, $\dfrac{4x^3}{16x^{-2}} = \dfrac{1}{4}x^5$.

Caution—Major Mistake Territory!

You cannot divide exponential expressions with different bases.

$5^2 \div 8^3$ cannot be simplified, because 5^2 and 8^3 do not have the same base.

$\dfrac{a^4}{b^5}$ cannot be simplified, because a^4 and b^5 do not have the same base.

BRAIN TICKLERS
Set # 28

Simplify the following expressions.

1. $\dfrac{2^3}{2^1}$

2. $\dfrac{2^4}{2^{-2}}$

3. $\dfrac{2x^5}{x^5}$

4. $\dfrac{2a^{-2}}{4a^2}$

5. $\dfrac{3x^4}{2x^{-7}}$

(Answers are on page 160.)

RAISING TO A POWER

When you raise an exponential expression to a power, multiply the exponents. Read each of the following examples carefully. Each example illustrates something different about simplifying exponents.

Simplify $(5^3)^2$.
When you raise an exponential expression to a power, multiply the exponents.
So, $(5^3)^2 = 5^{(3)(2)} = 5^6$.

Simplify $(3^4)^5$.
When you raise an exponential expression to a power, multiply the exponents.
So, $(3^4)^5 = 3^{(4)(5)} = 3^{20}$.

Simplify $(2^2)^{-3}$.
When you raise an exponential expression to a power, multiply the exponents. Remember, a positive number times a negative number is a negative number.
So, $(2^2)^{-3} = 2^{(2)(-3)} = 2^{-6}$.

Simplify $(7^{-3})^{-6}$.
Multiply the exponents. Remember, a negative number times a negative number is a positive number.
So, $(7^{-3})^{-6} = 7^{(-3)(-6)} = 7^{18}$.

Simplify $(8^4)^0$.
Multiply the exponents. Remember, any number times zero is zero.
So, $(8^4)^0 = 8^{(4)(0)} = 8^0 = 1$.

BRAIN TICKLERS
Set # 29

Simplify each of the following exponential expressions.

1. $(5^2)^5$

2. $(5^3)^{-1}$

3. $(5^{-2})^{-2}$

4. $(5^4)^0$

5. $(5^2)^3$

(Answers are on page 160.)

NEGATIVE EXPONENTS

Sometimes exponential expressions have negative exponents, such as 5^{-2}, x^{-3}, $2x^{-1}$, and 6^{-2}. In order to solve problems with negative exponents, you have to find the reciprocal of a number.

When you take the reciprocal of a number, you make the numerator the denominator and the denominator the numerator. It's just like turning a fraction upside down.

Watch, it's easy.

The reciprocal of $\frac{1}{2}$ is $\frac{2}{1}$.

The reciprocal of $\frac{5}{3}$ is $\frac{3}{5}$.

The reciprocal of $-\frac{3}{4}$ is $-\frac{4}{3}$.

What happens when you want to take the reciprocal of a whole number? First change the whole number to a fraction, and then take the reciprocal of that improper fraction.

What is the reciprocal of 5? Because 5 is $\frac{5}{1}$, the reciprocal of 5 is $\frac{1}{5}$.

What is the reciprocal of 6? Because 6 is $\frac{6}{1}$, the reciprocal of 6 is $\frac{1}{6}$.

What is the reciprocal of -3? Because -3 is $-\frac{3}{1}$, the reciprocal of -3 is $-\frac{1}{3}$.

What happens when you want to take the reciprocal of a mixed number?

Change the mixed number to an improper fraction and take the reciprocal of that fraction. It's *painless*.

What is the reciprocal of $5\frac{1}{2}$?

Change $5\frac{1}{2}$ into $\frac{11}{2}$.

Take the reciprocal of $\frac{11}{2}$.

The reciprocal of $\frac{11}{2}$ is $\frac{2}{11}$.

So, the reciprocal of $5\frac{1}{2}$ is $\frac{2}{11}$.

What is the reciprocal of $-6\frac{2}{3}$?

Change $-6\frac{2}{3}$ to $-\frac{20}{3}$.

Take the reciprocal of $-\frac{20}{3}$.

The reciprocal of $-\frac{20}{3}$ is $-\frac{3}{20}$.

So, the reciprocal of $-6\frac{2}{3}$ is $-\frac{3}{20}$.

BRAIN TICKLERS
Set # 30

Find the reciprocal of the following numbers.

1. 3

2. -8

3. $\frac{4}{3}$

4. $-\frac{2}{3}$

5. $8\frac{1}{2}$

6. $2x$

7. $x - 1$

8. $-\frac{x}{3}$

(Answers are on page 161.)

Changing a negative exponent to a positive exponent is a two-step process.

Step 1: Take the reciprocal of the number that is raised to a power.

Step 2: Change the exponent from a negative one to a positive one.

Simplify 5^{-3}.

Step 1: The reciprocal of five is $\frac{1}{5}$.

Step 2: Change the exponent from negative three to positive three.
So, $5^{-3} = \frac{1}{5^3}$.

Simplify 2^{-1}.

Step 1: The reciprocal of two is $\frac{1}{2}$.

Step 2: Change the exponent from negative one to positive one.
So, $2^{-1} = \frac{1}{2^1} = \frac{1}{2}$.

Simplify x^{-3}.

Step 1: The reciprocal of x is $\frac{1}{x}$.

Step 2: Change the exponent from negative three to positive three.
So, $x^{-3} = \frac{1}{x^3}$.

Simplify $(x - 2)^{-4}$.

Step 1: The reciprocal of $(x - 2)$ is $\frac{1}{(x - 2)}$.

Step 2: Change the exponent from negative four to positive four.
So, $(x - 2)^{-4} = \frac{1}{(x - 2)^4}$.

Simplify $\left(\frac{3}{5}\right)^{-2}$.

Step 1: The reciprocal of $\frac{3}{5}$ is $\frac{5}{3}$.

Step 2: Change the exponent from negative two to positive two.
So, $\left(\frac{3}{5}\right)^{-2} = \left(\frac{5}{3}\right)^2$.

BRAIN TICKLERS
Set # 31

Change the following negative exponents to positive exponents.

1. 4^{-3}

2. 3^{-4}

3. 2^{-5}

4. $\left(\frac{2}{5}\right)^{-2}$

5. $\left(\frac{1}{x}\right)^{-3}$

6. $\left(6\frac{1}{2}\right)^{-1}$

7. $(x-4)^{-2}$

(Answers are on page 161.)

SUPER BRAIN TICKLERS

Compute the value of the following exponential expressions.

1. 3^3

2. $\left(\frac{1}{3}\right)^2$

3. 3^{-1}

4. $\left(\frac{1}{3}\right)^{-2}$

5. $(-3 \cdot 3)^1$

6. $3^2 \cdot 3^{-1}$

7. $\frac{3^4}{3^1}$

8. $(3^2)^2$

9. $3^2 + 3^1$

10. $(4-1)^{-3}$

(Answers are on page 162.)

WORD PROBLEMS

Watch how to solve these word problems using exponents.

PROBLEM 1: A number squared plus six squared equals one hundred. What is the number?

Change this problem from Plain English to Math Talk.

"A number squared" becomes "x^2."
"Plus" becomes "+."
"Six squared" becomes "6^2."
"Equals" becomes "=."
"One hundred" becomes "100."
The result is the equation $x^2 + 6^2 = 100$.

Solve this equation. First, square six.
$6^2 = 36$
The result is the equation $x^2 + 36 = 100$.
Next, subtract 36 from both sides of the equation.
$x^2 + 36 - 36 = 100 - 36$

Compute.
$x^2 = 64$
Take the square root of both sides of this equation.
$x = 8$

Check your answer.
$8^2 + 6^2 = 100$
$64 + 36 = 100$
The solution is correct.

PROBLEM 2: Four times a number squared minus that number squared is seventy-five. What is that number?

Change this problem from Plain English to Math Talk.

"Four times a number squared" becomes "$4x^2$."
"Minus" becomes "−."
"That number squared" becomes "x^2."
"Is" becomes "=."
"Seventy-five" becomes "75."
The result is the equation $4x^2 - x^2 = 75$.

Solve this equation. First subtract.

$4x^2 - x^2 = 3x^2$

You can subtract because both of these expressions have the same base and the same exponent.

The result is the equation $3x^2 = 75$.

Divide both sides of the equation by three.

$$\frac{3x^2}{3} = \frac{75}{3}$$

Compute.

$x^2 = 25$

Take the square root of both sides of the equation.

$x = 5$

Check the answer.

$$4(5^2) - (5^2) = 75$$
$$4(25) - 25 = 75$$
$$100 - 25 = 75$$
$$75 = 75$$

The solution is correct.

BRAIN TICKLERS— THE ANSWERS

Set # 23, page 134

1. D

2. C

3. E

4. F

5. A

6. B

Set # 24, page 137

1. $5^2 = 25$

2. $2^6 = 64$

3. $10^2 = 100$

4. $4^3 = 64$

5. $5^0 = 1$

6. $(-3)^2 = 9$

7. $(-5)^3 = -125$

8. $(-4)^2 = 16$

9. $(-1)^5 = -1$

10. $(-1)^{12} = 1$

Set # 25, page 140

1. $3(5)^2 = 3(25) = 75$

2. $-4(3)^2 = -4(9) = -36$

3. $2(-1)^2 = 2(1) = 2$

4. $3(-1)^3 = 3(-1) = -3$

5. $5(-2)^2 = 5(4) = 20$

6. $-\frac{1}{2}(-4)^2 = -\frac{1}{2}(16) = -8$

7. $-2(-3)^2 = -2(9) = -18$

8. $-3(-3)^3 = -3(-27) = 81$

Set # 26, page 143

1. $3(3)^2 + 5(3)^2 = 8(3)^2 = 72$

2. $4(16)^3 - 2(16)^3 = 2(16)^3 = 8192$

3. $3x^2 - 5x^2 = -2x^2$

4. $2x^0 + 5x^0 = 7x^0 = 7$

5. $5x^4 - 5x^4 = 0x^4 = 0$

Set # 27, page 146

1. $2^3 2^3 = 2^6$

2. $2^5 2^2 = 2^7$

3. $2^{10} 2^{-2} = 2^8$

4. $2^{-1} \cdot 2^3 \cdot 2^1 = 2^3$

5. $x^3 x^{-2} = x^1$

6. $x^4 \cdot x^{-4} = x^0$

7. $6x^4(x^{-2}) = 6x^2$

8. $-7x^2(5x^3) = -35x^5$

9. $(-6x^3)(-2x^{-3}) = 12x^0 = 12(1) = 12$

Presto! You've got it!

Set # 28, page 149

1. $\dfrac{2^3}{2^1} = 2^2 = 4$

2. $\dfrac{2^4}{2^{-2}} = 2^6 = 64$

3. $\dfrac{2x^5}{x^5} = 2$

4. $\dfrac{2a^{-2}}{4a^2} = \dfrac{1}{2} a^{-4}$

5. $\dfrac{3x^4}{2x^{-7}} = \dfrac{3}{2} x^{11}$

Set # 29, page 151

1. $(5^2)^5 = 5^{10}$

2. $(5^3)^{-1} = 5^{-3}$

3. $(5^{-2})^{-2} = 5^4$

4. $(5^4)^0 = 5^0$

5. $(5^2)^3 = 5^6$

Set # 30, page 153

1. The reciprocal of 3 is $\frac{1}{3}$.

2. The reciprocal of -8 is $-\frac{1}{8}$.

3. The reciprocal of $\frac{4}{3}$ is $\frac{3}{4}$.

4. The reciprocal of $-\frac{2}{3}$ is $-\frac{3}{2}$.

5. The reciprocal of $8\frac{1}{2}$ is $\frac{2}{17}$.

6. The reciprocal of $2x$ is $\frac{1}{2x}$.

7. The reciprocal of $x - 1$ is $\frac{1}{x - 1}$.

8. The reciprocal of $-\frac{x}{3}$ is $-\frac{3}{x}$.

Set # 31, page 155

1. $4^{-3} = \left(\frac{1}{4}\right)^3$

2. $3^{-4} = \left(\frac{1}{3}\right)^4$

3. $2^{-5} = \left(\frac{1}{2}\right)^5$

4. $\left(\frac{2}{5}\right)^{-2} = \left(\frac{5}{2}\right)^2$

5. $\left(\frac{1}{x}\right)^{-3} = x^3$

6. $\left(6\frac{1}{2}\right)^{-1} = \left(\frac{2}{13}\right)^1$

7. $(x - 4)^{-2} = \left(\frac{1}{x - 4}\right)^2$

Super Brain Ticklers, page 155

1. $3^3 = 27$

2. $\left(\frac{1}{3}\right)^2 = \frac{1}{9}$

3. $3^{-1} = \frac{1}{3}$

4. $\left(\frac{1}{3}\right)^{-2} = 9$

5. $(-3 \cdot 3)^1 = -9$

6. $3^2 \cdot 3^{-1} = 3$

7. $\frac{3^4}{3^1} = 27$

8. $(3^2)^2 = 81$

9. $3^2 + 3^1 = 12$

10. $(4 - 1)^{-3} = \frac{1}{27}$

CHAPTER SIX

Roots and Radicals

SQUARE ROOTS

When you square a number, you multiply a number by itself. When you take the square root of a number, you try and figure out, "What number if multiplied by itself will give me this number?"

For example, to figure out, "What is the square root of twenty-five?," rephrase the question as, "What number, when multiplied by itself, equals twenty-five?" The answer is five. Five times five equals twenty-five. Five is the square root of twenty-five.

To figure out, "What is the square root of one hundred?," think of the question as, "What number when multiplied by itself will equal one hundred?" The answer is ten. Ten times ten equals one hundred. Ten is the square root of one hundred.

You write a square root by putting a number under a *radical sign*. A radical sign looks like this: $\sqrt{}$. When you see a radical sign, take the square root of the number under it. The number under the radical sign is called a *radicand*. Look at the problem $\sqrt{5}$. Five is under the radical sign. Five is the radicand.

MATH TALK!

Watch as the following expressions are
changed from Math Talk into Plain English.

$$\sqrt{9}$$

What is the square root of nine?
What number times itself equals nine?

$$\sqrt{16}$$

What is the square root of sixteen?
What number times itself equals sixteen?

When you figure out the square root of a number, you can
check your answer by multiplying the number by itself and see-
ing if the answer is the radicand. For example, what is the
square root of nine? The square root of nine is three. To check,
multiply three times three. Three times three is nine. The answer
is correct.

Certain numbers are *perfect squares*. The square root of a
perfect square is a whole number. For example, 16 and 25 are
both perfect squares.

That's a perfect square!

Every number has both a positive and a negative square root. When the square root is written without any sign in front of it, the answer is the positive square root. When the square root is written with a negative sign in front of it, the answer is the negative square root.

16 is a perfect square.
16 has both a negative and positive square root.
$\sqrt{16}$ is 4, since $4(4) = 16$.
The negative square root of 16 is -4, since $(-4)(-4) = 16$.

25 is a perfect square.
25 has both a negative and a positive square root.
$\sqrt{25}$ is 5, since $(5)(5) = 25$.
The negative square root of 25 is -5, since $(-5)(-5) = 25$.

Perfect Square Roots

Memorize all of these.

What is $\sqrt{0}$?	It is 0.	Check it. Square 0.	$0^2 = 0 \cdot 0 = 0$
What is $\sqrt{1}$?	It is 1.	Check it. Square 1.	$1^2 = 1 \cdot 1 = 1$
What is $\sqrt{4}$?	It is 2.	Check it. Square 2.	$2^2 = 2 \cdot 2 = 4$
What is $\sqrt{9}$?	It is 3.	Check it. Square 3.	$3^2 = 3 \cdot 3 = 9$
What is $\sqrt{16}$?	It is 4.	Check it. Square 4.	$4^2 = 4 \cdot 4 = 16$
What is $\sqrt{25}$?	It is 5.	Check it. Square 5.	$5^2 = 5 \cdot 5 = 25$
What is $\sqrt{36}$?	It is 6.	Check it. Square 6.	$6^2 = 6 \cdot 6 = 36$
What is $\sqrt{49}$?	It is 7.	Check it. Square 7.	$7^2 = 7 \cdot 7 = 49$
What is $\sqrt{64}$?	It is 8.	Check it. Square 8.	$8^2 = 8 \cdot 8 = 64$
What is $\sqrt{81}$?	It is 9.	Check it. Square 9.	$9^2 = 9 \cdot 9 = 81$
What is $\sqrt{100}$?	It is 10.	Check it. Square 10.	$10^2 = 10 \cdot 10 = 100$

BRAIN TICKLERS
Set # 32

Quickly solve these square root problems.

___ 1. $\sqrt{9}$ ___ 6. $\sqrt{100}$

___ 2. $\sqrt{25}$ ___ 7. $\sqrt{16}$

___ 3. $\sqrt{36}$ ___ 8. $\sqrt{4}$

___ 4. $\sqrt{81}$ ___ 9. $\sqrt{64}$

___ 5. $\sqrt{49}$ ___ 10. $\sqrt{1}$

(Answers are on page 200.)

The radicands in the previous Brain Ticklers were all perfect squares. Some numbers are not perfect squares. If a number is not a perfect square, no *whole* number when multiplied by itself will equal that number.

5 is not a perfect square; $\sqrt{5}$ is not a whole number.
3 is not a perfect square; $\sqrt{3}$ is not a whole number.

CUBE ROOTS AND HIGHER

Hey! I need that to find my root!!

When a number is multiplied by itself three times, it is *cubed*. Two cubed is two times two times two. Two cubed is eight. The cube root of a number is the opposite of cubing a number. "What is the cube root of eight?" is really asking the question, "What number when multiplied by itself three times equals eight?" The answer is two. Two times two times two is equal to eight. Two is the cube root of eight.

To write a *cube root*, draw a radical sign and put a 3 inside the crook of the radical sign. The expression $\sqrt[3]{8}$ is read, "What is the cube root of eight?" The number 3 is called the *index*. It tells you how many times the number must be multiplied by itself.

Notice the difference between these four sentences and their mathematical equivalents.

The cube root of eight is two. $\qquad \sqrt[3]{8} = 2$
Two is the cube root of eight. $\qquad 2 = \sqrt[3]{8}$
Two cubed is eight. $\qquad\qquad\quad 2^3 = 8$
Eight is two cubed. $\qquad\qquad\quad\; 8 = 2^3$

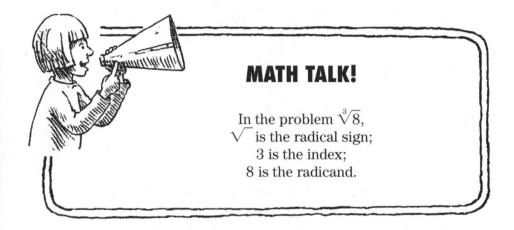

MATH TALK!

In the problem $\sqrt[3]{8}$,
$\sqrt{}$ is the radical sign;
3 is the index;
8 is the radicand.

It is also possible to take the fourth root of a number. To write, "What is the fourth root of a number?" in mathematical terms, write four as the index in the radical sign: $\sqrt[4]{}$. The radical expression now poses the question, "What number when multiplied by itself four times equals the number under the radical sign?" The radical expression $\sqrt[4]{16}$ is read "What is the fourth root of sixteen?" The problem can be rephrased as, "What number when multiplied by itself four times is equal to sixteen?" The answer is two. Two times two is four, times two is eight, times two is sixteen.

$$2^4 = 16 \qquad \text{Two to the fourth power is sixteen.}$$
$$\sqrt[4]{16} = 2 \qquad \text{The fourth root of sixteen is two.}$$

The index of a radical can also be a natural number greater than 4. If the index is ten, it asks, "What number multiplied by itself ten times is equal to the number under the radical sign?" If the index is 100, it asks, "What number when multiplied by itself 100 times is equal to the number under the radical sign?"

MATH TALK!

The following radical expressions are changed from Math Talk into Plain English.

$$\sqrt[3]{27} = 3$$

The cube root of twenty-seven is three.

$$3 = \sqrt[3]{27}$$

Three is the cube root of twenty-seven.

$$\sqrt[4]{16} = 2$$

The fourth root of sixteen is two.

$$2 = \sqrt[4]{16}$$

Two is the fourth root of sixteen.

$$\sqrt[6]{64} = 2$$

The sixth root of sixty-four is two.

$$2 = \sqrt[6]{64}$$

Two is the sixth root of sixty-four.

BRAIN TICKLERS
Set # 33

Solve the following radical expressions.

___ 1 $\sqrt[3]{27}$

___ 4. $\sqrt[4]{16}$

___ 2. $\sqrt[3]{64}$

___ 5. $\sqrt[3]{125}$

___ 3. $\sqrt[5]{1}$

___ 6. $\sqrt[10]{0}$

(Answers are on page 200.)

Sometimes the index of a radical is a variable.
For example, $\sqrt[x]{9}$, $\sqrt[y]{16}$, $\sqrt[x]{25}$, $\sqrt[y]{32}$.

MATH TALK!

Watch as these radical expressions are changed
from Math Talk into Plain English.

$$\sqrt[x]{9}$$

What is the xth root of nine?
What number multiplied by itself
x times equals nine?

$$\sqrt[y]{16}$$

What is the yth root of sixteen?
What number multiplied by itself y times
equals sixteen?

It is impossible to compute the value of a radical expression when the index is a variable. For example, it is impossible to compute the value of $\sqrt[y]{16}$, unless you know the value of y. If $y = 2$, the value of $\sqrt[y]{16}$ is 4. But if $y = 4$, the value of $\sqrt[y]{16}$ is 2.

It is possible to solve for the index if you know the value of the radical expression. To solve $\sqrt[x]{9} = 3$ for x, rewrite this radical expression as an exponential expression.

Rewrite $\sqrt[x]{9} = 3$ as $3^x = 9$.

Solve for x by substituting different numbers for x.
If $x = 1$, then $3^x = 3^1 = 3$.
If $x = 2$, $3^x = 3^2 = 9$.
So, $x = 2$.

Solve for x: $\sqrt[x]{125} = 5$

Rewrite $\sqrt[x]{125}$ as the exponential expression $5^x = 125$.
Substitute different natural numbers for x. Start with one.
If $x = 1$, $5^x = 5^1 = 5$; x is not equal to one.
If $x = 2$, $5^x = 5^2 = 25$; x is not equal to two.
If $x = 3$, $5^x = 5^3 = 125$.
So, $x = 3$.

BRAIN TICKLERS
Set # 34

Solve for x.

_____ 1. $\sqrt[x]{16} = 4$ _____ 4. $\sqrt[x]{125} = 5$

_____ 2. $\sqrt[x]{16} = 2$ _____ 5. $\sqrt[x]{8} = 2$

_____ 3. $\sqrt[x]{25} = 5$

(Answers are on page 201.)

NEGATIVE RADICANDS

What happens when the number under the radical sign is a negative number?

For example:

$\sqrt{-4}$ What is the square root of negative four?

$\sqrt[3]{-8}$ What is the cube root of negative eight?

$\sqrt[4]{-16}$ What is the fourth root of negative sixteen?

$\sqrt[5]{-32}$ What is the fifth root of negative thirty-two?

The answer depends on whether the index is even or the index is odd.

Case 1: The index is even.

When the index is even, you cannot compute the value of a negative radicand. Watch.

What is $\sqrt{-4}$?

Try to find the square root of negative four.

What number when multiplied by itself equals negative four?

$$(+2)(+2) = +4, \text{not} -4.$$
$$(-2)(-2) = +4, \text{not} -4.$$

There is no real number that when multiplied by itself will equal negative four.

What is $\sqrt[4]{-16}$?

$$(+2)(+2)(+2)(+2) = +16, \text{not} -16.$$
$$(-2)(-2)(-2)(-2) = +16, \text{not} -16.$$

There is no real number that when multiplied by itself four times is negative sixteen.

The even root of a negative number is always undefined in the real number system.

$$\sqrt[4]{-81} \text{ is undefined.}$$
$$\sqrt{-25} \text{ is undefined.}$$
$$\sqrt[10]{-20} \text{ is undefined.}$$
$$\sqrt[8]{-100} \text{ is undefined.}$$
$$\sqrt[100]{-1} \text{ is undefined.}$$

Caution—Major Mistake Territory!

$\sqrt{-9}$ is not negative three.

$\sqrt[4]{-16}$ is not negative two.

When the index of a radical is even, you cannot take the root of a negative number.

Case 2: The index is odd.

What is $\sqrt[3]{-8}$?

The expression $\sqrt[3]{-8}$ asks, "What number when multiplied by itself three times is equal to negative eight?" The answer is negative two.

$$(-2)(-2)(-2) = -8$$

Negative two times negative two times negative two is equal to negative eight. Why?

Multiply the first two negative twos together.

$$(-2)(-2) = 4$$

Negative two times negative two equals positive four. Now multiply four times the last negative two.

$$(4)(-2) = (-8)$$

Four times negative two is negative eight.

When you cube a negative number, the answer is negative.

$$\sqrt[3]{-8} = -2$$

What is $\sqrt[5]{-32}$?

What is the fifth root of negative thirty-two? What number when multiplied by itself five times equals negative thirty-two? The answer is negative two.

$$(-2)(-2)(-2)(-2)(-2) = -32$$

Negative two times two is positive four, times negative two is negative eight, times negative two is positive sixteen, times negative two is negative thirty-two.

$$\sqrt[5]{-32} = -2$$

If the index of a radical is odd, the problem has a solution. If the number under the radical sign is positive, the solution is positive. If the number under the radical sign is negative, the solution is negative.

Watch carefully what happens when -1 is under the radical sign.

$$\sqrt{-1} = \text{undefined}$$
$$\sqrt[3]{-1} = -1$$
$$\sqrt[7]{-1} = -1$$
$$\sqrt[10]{-1} = \text{undefined}$$
$$\sqrt[25]{-1} = -1$$
$$\sqrt[100]{-1} = \text{undefined}$$

Caution—Major Mistake Territory!

The square root of a negative number is not a negative number.
The square root of a negative number is undefined.

$$\sqrt{-9} \text{ is undefined.}$$
$$\sqrt{-25} \text{ is undefined.}$$
$$\sqrt{-36} \text{ is undefined.}$$

BRAIN TICKLERS
Set # 35

Solve the following radicals. Be careful. Some of them are undefined.

_____ 1. $\sqrt{-64}$ _____ 4. $\sqrt[5]{-32}$

_____ 2. $\sqrt[4]{-16}$ _____ 5. $\sqrt{-49}$

_____ 3. $\sqrt[3]{-27}$ _____ 6. $\sqrt[9]{-1}$

(Answers are on page 201.)

RADICAL EXPRESSIONS

Simplifying radical expressions

There are rules for simplifying radical expressions.

RULE 1: If two numbers are multiplied under a radical sign, you can rewrite them under two different radical signs that are multiplied together.

$$\sqrt{(9)(16)} = (\sqrt{9})(\sqrt{16})$$
$$\sqrt{(25)(4)} = (\sqrt{25})(\sqrt{4})$$
$$\sqrt{(36)(100)} = (\sqrt{36})(\sqrt{100})$$

Watch how separating a radical expression into two radical expressions makes simplifying easy.

Simplify $\sqrt{(9)(16)}$.
Rewrite $\sqrt{(9)(16)}$ as two separate expressions.
$$\sqrt{(9)(16)} = (\sqrt{9})(\sqrt{16})$$
Solve each of the two radical expressions.
$$(\sqrt{9}) = 3 \text{ and } (\sqrt{16}) = 4$$
Multiply the solutions together.
$$(3)(4) = 12$$
The result is the answer.
$$\sqrt{(9)(16)} = 12$$

Simplify $\sqrt{(25)(4)}$.
Rewrite $\sqrt{(25)(4)}$ as two separate expressions.
$$\sqrt{(25)(4)} = (\sqrt{25})\sqrt{(4)}.$$
Solve each of the two radical expressions.
$$(\sqrt{25}) = 5 \text{ and } (\sqrt{4}) = 2$$
Multiply the solutions together.
$$(5)(2) = 10$$
The result is the answer.
$$\sqrt{(25)(4)} = 10$$

Simplify $\sqrt{36x^2}$. ($x > 0$)

Rewrite $36x^2$ as two separate expressions.
$$\sqrt{36x^2} = \sqrt{36}\ \sqrt{x^2}$$

Solve each of the two radical expressions.
$$\sqrt{36} = 6 \text{ and } \sqrt{x^2} = x$$

Multiply the solutions together.
$$(6)(x) = 6x$$

The result is the answer.
$$\sqrt{36x^2} = 6x$$

Simplify $\sqrt{(4)(16)(25)}$.

Rewrite $\sqrt{(4)(16)(25)}$ as three separate expressions.
$$\sqrt{(4)(16)(25)} = (\sqrt{4})(\sqrt{16})(\sqrt{25})$$

Solve each of the three radical expressions.
$$\sqrt{4} = 2 \text{ and } \sqrt{16} = 4 \text{ and } \sqrt{25} = 5$$

Multiply the solutions together.
$$(2)(4)(5) = 40$$

The result is the answer.
$$\sqrt{(4)(16)(25)} = 40$$

Sometimes both of the numbers under the radical sign are not perfect squares. Take the square root of the numbers that are perfect squares and multiply the result by the square root of the number that is not a perfect square. Does it sound complicated? It's not. It's *painless*.

Simplify $\sqrt{(25)(6)}$.

Rewrite $\sqrt{(25)(6)}$ as two separate expressions.
$$\sqrt{(25)(6)} = (\sqrt{25})(\sqrt{6})$$

Take the square root of the expressions that are perfect squares.
$$\sqrt{25} = 5; \qquad \sqrt{6} \text{ is not a perfect square.}$$

Multiply the solutions together.
$$5\sqrt{6}$$

The result is the answer.
$$\sqrt{(25)(6)} = 5\sqrt{6}$$
It isn't necessary to find the actual value of $\sqrt{6}$.

Simplify $\sqrt{49x}$.
Rewrite $\sqrt{49x}$ as two separate expressions.
$$\sqrt{49x} = \sqrt{49} \ \sqrt{x}$$
Take the square root of the expressions that are perfect squares.
$$\sqrt{49} = 7; \qquad \sqrt{x} \text{ is not a perfect square.}$$
Multiply the two solutions together.
$$7\sqrt{x}$$
The result is the answer.
$$\sqrt{49x} = 7\sqrt{x}$$

BRAIN TICKLERS
Set # 36

Solve the following square root problems by rewriting them as two separate radical expressions.

1. $\sqrt{(16)(9)}$

2. $\sqrt{(64)(100)}$

3. $\sqrt{25y^2}\ (y > 0)$

4. $\sqrt{(4)(11)}$

5. $\sqrt{9y}$

(Answers are on page 201.)

RULE 2: If two radical expressions are multiplied together, you rewrite them as products under the same radical sign.

$(\sqrt{27})(\sqrt{3})$ can be rewritten as $\sqrt{(27)(3)}$.
$(\sqrt{5})(\sqrt{5})$ can be rewritten as $\sqrt{(5)(5)}$.
$(\sqrt{8})(\sqrt{5})$ can be rewritten as $\sqrt{(8)(5)}$.

Rewriting two radical expressions that are multiplied together under the same radical sign can often help simplify them. Watch.

Simplify $(\sqrt{27})(\sqrt{3})$.
It is impossible to simplify $\sqrt{27}$ or $\sqrt{3}$.
Put both of these radical expressions under the same radical sign.
$$(\sqrt{27})(\sqrt{3}) = \sqrt{(27)(3)}$$
Multiply the expression under the radical sign.
$$\sqrt{(27)(3)} = \sqrt{81}$$
81 is a perfect square.
$$\sqrt{81} = 9$$
The result is the answer.
$$(\sqrt{27})(\sqrt{3}) = 9$$

Simplify $(\sqrt{5})(\sqrt{5})$.
It is impossible to simplify each $\sqrt{5}$.
Put both of these radical expressions under the same radical sign.
$$(\sqrt{5})(\sqrt{5}) = \sqrt{(5)(5)}$$
Multiply the expression under the radical sign.
$$\sqrt{(5)(5)} = \sqrt{25}$$
25 is a perfect square.
$$\sqrt{25} = 5$$
The result is the solution.
$$(\sqrt{5})(\sqrt{5}) = 5$$

BRAIN TICKLERS
Set # 37

Simplify these radical expressions by placing them under the same radical sign.

1. $(\sqrt{3})(\sqrt{3})$

2. $(\sqrt{8})(\sqrt{2})$

3. $(\sqrt{12})(\sqrt{3})$

4. $(\sqrt{x})(\sqrt{x})$ $(x > 0)$

5 $(\sqrt{x^3})(\sqrt{x})$ $(x > 0)$

(Answers are on page 202.)

Factoring a radical expression

There must be a way to simplify this square root!

Sometimes the number under the square root sign is not a perfect square, but it can still be simplified.

RULE 3: You can factor the number under a radical and take the square root of the factors.

For example, what is $\sqrt{12}$?
What number when multiplied by itself is equal to twelve? There is no whole number that when multiplied by itself will equal twelve. If you use a calculator and enter the number twelve, and tap the square root symbol, 3.464101615138 will show up on the screen.
So, (3.464101615138)(3.464101615138) = 12.

But besides using a calculator to compute square roots, mathematicians often simplify square roots. To simplify a square root, look at the factors of the number under the radical sign.

Simplify $\sqrt{12}$.
What are the factors of 12?
$$（3)(4) = 12$$
$$(2)(6) = 12$$
Are any of the factors perfect squares?
Out of all of these numbers only (4) is a perfect square.
$$(2)(2) = 4$$
Rewrite $\sqrt{12}$ as $\sqrt{(4)(3)}$.
Rewrite $\sqrt{(4)(3)}$ as $(\sqrt{4})(\sqrt{3})$.
Now $\sqrt{4} = 2$, so rewrite $(\sqrt{4})(\sqrt{3})$ as $2\sqrt{3}$.
So, $\sqrt{12}$ is $2\sqrt{3}$.

Simplify $\sqrt{18}$.
Find the factors of 18.
$$(9)(2) = 18$$
$$(3)(6) = 18$$
Are any of the factors perfect squares?
Nine is a perfect square, so $\sqrt{18}$ can be simplified.
Rewrite $\sqrt{18}$ as $(\sqrt{9})(\sqrt{2})$.
Now $\sqrt{9} = 3$, so rewrite $(\sqrt{9})(\sqrt{2})$ as $3\sqrt{2}$.
So, $\sqrt{18}$ is $3\sqrt{2}$.

BRAIN TICKLERS
Set # 38

Simplify the following radical expressions by factoring.

1. $\sqrt{20}$ 4. $\sqrt{24}$

2. $\sqrt{8}$ 5. $\sqrt{32}$

3. $\sqrt{27}$ 6. $\sqrt{125}$

(Answers are on page 202.)

Division of radicals

RULE 4: If two numbers are divided under a radical sign, you can rewrite them under two different radical signs separated by a division sign.

$\sqrt{\dfrac{9}{4}}$ can be rewritten as $\dfrac{\sqrt{9}}{\sqrt{4}}$

$\sqrt{\dfrac{25}{36}}$ can be rewritten as $\dfrac{\sqrt{25}}{\sqrt{36}}$

$\sqrt{\dfrac{64}{16}}$ can be rewritten as $\dfrac{\sqrt{64}}{\sqrt{16}}$.

To take the square root of a rational number, rewrite the problem and put the numerator and denominator under two different radical signs. Next take the square root of each one of the numbers. Watch. It's *painless*.

Simplify $\sqrt{\dfrac{9}{4}}$.
Rewrite the problem as $\dfrac{\sqrt{9}}{\sqrt{4}}$.
Solve as $\sqrt{9}$ and as $\sqrt{4}$.

$$\sqrt{9} = 3 \text{ and } \sqrt{4} = 2$$

So, $\sqrt{\dfrac{9}{4}} = \dfrac{3}{2}$.

Simplify $\sqrt{\frac{x^2}{16}}$.

Rewrite the problem as $\frac{\sqrt{x^2}}{\sqrt{16}}$.

In this example, $x > 0$.

Solve as $\sqrt{x^2}$ and as $\sqrt{16}$.

$$\sqrt{x^2} = x \text{ and } \sqrt{16} = 4$$

So, $\sqrt{\frac{x^2}{16}} = \frac{x}{4}$.

Simplify $\sqrt{\frac{3}{25}}$.

Rewrite the problem as $\frac{\sqrt{3}}{\sqrt{25}}$.

Solve as $\sqrt{3}$ and as $\sqrt{25}$.

$$\sqrt{3} \text{ cannot be simplified. } \sqrt{25} = 5$$

So, $\sqrt{\frac{3}{25}} - \frac{\sqrt{3}}{5}$.

Simplify $\sqrt{\frac{49}{5}}$.

Rewrite the problem as $\frac{\sqrt{49}}{\sqrt{5}}$.

Solve as $\sqrt{49}$ and as $\sqrt{5}$.

$$\sqrt{49} = 7; \ \sqrt{5} \text{ cannot be simplified.}$$

So, $\sqrt{\frac{49}{5}} = \frac{7}{\sqrt{5}}$.

This is the answer.

However, the answer has a radical in the denominator.

If an answer has a radical in the denominator it is not considered simplified. In order to simplify this expression you have to rationalize the denominator. Let's see how.

Rationalizing the denominator

A radical expression is not simplified if there is a radical expression in the denominator.

$\frac{3}{\sqrt{2}}$ is not simplified because $\sqrt{2}$ is in the denominator.

$\frac{\sqrt{5}}{\sqrt{3}}$ is not simplified because $\sqrt{3}$ is in the denominator.

$\frac{5}{\sqrt{x}}$ is not simplified because $\sqrt{x}$ is in the denominator.

RULE 5: You can multiply the numerator and denominator of a radical expression by the same number and not change the value of the expression.

This will help you eliminate radical expressions in the denominator.

To eliminate a radical expression in the denominator of a rational expression, follow these *painless* steps.

Step 1: Identify the radical expression in the denominator.

Step 2: Construct a fraction where this radical expression is both the numerator and denominator of this fraction.

Step 3: Multiply the original expression by the fraction.

Step 4: Put the two square roots in the denominator under the same radical.

Step 5: Take the square root of the number in the denominator. The result is the answer.

Does it sound complicated? Watch. It's *painless*.

Rationalize the denominator in the expression $\frac{3}{\sqrt{2}}$.

Step 1: Identify the radical expression in the denominator. The radical expression in the denominator is $\sqrt{2}$.

Step 2: Construct a fraction where this radical expression is both the numerator and denominator. The value of the fraction is 1.
$$\frac{\sqrt{2}}{\sqrt{2}} = 1$$

Step 3: Multiply the original expression by this new expression.
$$\left(\frac{3}{\sqrt{2}}\right)\left(\frac{\sqrt{2}}{\sqrt{2}}\right) = \frac{3\sqrt{2}}{\sqrt{2}\sqrt{2}}$$

Step 4: Put the two square roots in the denominator under the same radical.

$$\frac{3\sqrt{2}}{\sqrt{2}\sqrt{2}} = \frac{3\sqrt{2}}{\sqrt{4}}$$

Step 5: Take the square root of the number in the denominator.

$$\frac{3\sqrt{2}}{\sqrt{4}} = \frac{3\sqrt{2}}{2}$$

The result is the solution.

$$\frac{3}{\sqrt{2}} = \frac{3\sqrt{2}}{2}$$

Simplify $\sqrt{\frac{25}{3}}$.

Step 1: Identify the radical expression in the denominator.

$$\sqrt{\frac{25}{3}} = \frac{\sqrt{25}}{\sqrt{3}} = \frac{5}{\sqrt{3}}$$

$\sqrt{3}$ is the radical expression in the denominator.

Step 2: Construct a fraction where this radical expression is both the numerator and denominator of this fraction.

$$\frac{\sqrt{3}}{\sqrt{3}} = 1$$

You've got it now.

Step 3: Multiply the original expression by this fraction.

$$\frac{5}{\sqrt{3}} \cdot \frac{\sqrt{3}}{\sqrt{3}} = \frac{5\sqrt{3}}{\sqrt{3}\sqrt{3}}$$

Step 4: Put the two square roots in the denominator under the same radical.

$$\frac{5\sqrt{3}}{\sqrt{3}\sqrt{3}} = \frac{5\sqrt{3}}{\sqrt{9}}$$

Step 5: Take the square root of the number in the denominator. The denominator is $\sqrt{9}$.

$$\frac{5\sqrt{3}}{\sqrt{9}} = \frac{5\sqrt{3}}{3}$$

The answer is the solution.

$$\sqrt{\frac{25}{3}} = \frac{5\sqrt{3}}{3}$$

BRAIN TICKLERS
Set # 39

Rationalize the denominator in the following expressions.

1. $\dfrac{5}{\sqrt{3}}$

2. $\dfrac{10}{\sqrt{2}}$

3. $\sqrt{\dfrac{16}{3}}$

4. $\sqrt{\dfrac{36}{7}}$

(Answers are on page 203.)

Adding and subtracting radical expressions

RULE 6: You can add and subtract radical expressions if each index is the same and each number under the radical sign is the same. When you add radical expressions, add the coefficients.

$2\sqrt{3} + 3\sqrt{3}$

First check to make sure the radicals have the same index. There is no index written in either of these expressions, so the index in both of them is two.

Next check if both radical expressions have the same quantity under the radical sign. Both expressions have 3 under the radical sign.

Add $2\sqrt{3}$ and $3\sqrt{3}$ by adding the coefficients.

> The coefficient of $2\sqrt{3}$ is 2.
> The coefficient of $3\sqrt{3}$ is 3.
> Add the coefficients: $2 + 3 = 5$.

Bring over the radical expression, $\sqrt{3}$.
So, $2\sqrt{3} + 3\sqrt{3} = 5\sqrt{3}$.

$\sqrt[3]{5} + 5\sqrt[3]{5}$

First check to make sure the radicals have the same index. Three is the index in both expressions.

Next check to see if both radical expressions have the same quantity under the radical sign. Both expressions have the number 5 under the radical sign.

Since both expressions have the same index and the same expression under the radical sign, they can be added together. To add $\sqrt[3]{5}$ and $5\sqrt[3]{5}$, just add the coefficients.

> The coefficient of $\sqrt[3]{5}$ is 1.
> The coefficient of $5\sqrt[3]{5}$ is 5.
> The coefficients are 1 and 5: $1 + 5 = 6$.

Bring over the radical expression, $\sqrt[3]{5}$.
So, $\sqrt[3]{5} + 5\sqrt[3]{5} = 6\sqrt[3]{5}$.

$\sqrt{x} - 4\sqrt{x}$

First check to make sure the radicals have the same index. There is no index written in either of these expressions, so the index in both of them is two.

Next check if both radical expressions have the same quantity under the radical sign. Both expressions have an x under the radical sign, so one can be subtracted from the other.

To subtract $\sqrt{x} - 4\sqrt{x}$, just subtract the coefficients.

The coefficient of $\sqrt{x}$ is 1.
The coefficient of $4\sqrt{x}$ is 4.
Subtract the coefficients. $1 - 4 = -3$.

Bring over the radical expression, $\sqrt{x}$.
So, $\sqrt{x} - 4\sqrt{x} = -3\sqrt{x}$.

Caution—Major Mistake Territory!

Radicals without the same index cannot be added or subtracted.

Radicals without the same number under the radical cannot be added or subtracted.

$\sqrt{x}$ and $\sqrt{2x}$ cannot be added or subtracted—
$\sqrt{x}$ and $\sqrt{2x}$ both have the same index,
but $\sqrt{x}$ and $\sqrt{2x}$ do not have the same expression under the radical sign.

$\sqrt[3]{7}$ and $\sqrt[4]{7}$ cannot be added or subtracted—
$\sqrt[3]{7}$ and $\sqrt[4]{7}$ do not have the same index, although
$\sqrt[3]{7}$ and $\sqrt[4]{7}$ have the same number under the radical sign.

BRAIN TICKLERS
Set # 40

Add and subtract the following radical expressions.

1. $7\sqrt{2} + \sqrt{2}$

2. $3\sqrt{3} + \sqrt{3}$

3. $5\sqrt[4]{x} + 2\sqrt[4]{x}$

4. $9\sqrt{5} - \sqrt{5}$

5. $2\sqrt{x} - 2\sqrt{x}$

6. $5\sqrt{2x} - 2\sqrt{2x}$

(Answers are on page 203.)

Fractional exponents

RULE 7: Radical expressions can be written as fractional exponents. The numerator of the exponent is the power of the radicand. The denominator of the exponent is the index. Watch.

$\sqrt[3]{x^2}$ is the same as $x^{\frac{2}{3}}$.

Follow these simple steps to change a radical expression to an exponential one.

Step 1: Copy the base under the radical sign. Put it in parentheses.

Step 2: Make the exponent the numerator of the expression.

Step 3: Make the index the denominator of the expression.

Change $\sqrt[3]{2^5}$ to an exponential expression.

Step 1: Copy the number or expression under the radical sign and put it in parentheses.
Two is the base of the radicand. Copy the two and put it in parentheses.
$$(2)$$

Step 2: Make the exponent the numerator of the expression.
Five is the exponent. Make it the numerator.
$$(2)^{\frac{5}{?}}$$

Step 3: Make the index the denominator of the expression.
Three is the index. Make three the denominator.
$$(2)^{\frac{5}{3}}$$
Answer: $\sqrt[3]{2^5} = (2)^{\frac{5}{3}}$

Change $\sqrt{5}$ to an exponential expression.

Step 1: Copy the number or expression under the radical sign. Put the entire expression in parentheses. Copy the five.

$$(5)$$

Step 2: Make the exponent of the radicand the numerator of the expression. The exponent is one. Make one the numerator.

$$(5)^{\frac{1}{?}}$$

Step 3: Make the index the denominator of the expression. Since there is no index written, the index is two. Make two the denominator.

$$(5)^{\frac{1}{2}}$$

Answer: $\sqrt{5} = (5)^{\frac{1}{2}}$

You can do it!

Change $\sqrt{(3xy)^3}$ to an exponential expression.

Step 1: Copy the number or expression under the radical sign. Put the entire expression in parentheses. Put the expression $3xy$ in parentheses.

$$(3xy)$$

Step 2: Make the exponent the numerator of the expression. Three is the exponent. Make it the numerator of the expression.

$$(3xy)^{\frac{3}{?}}$$

Step 3: Make the index the denominator of the expression. The index is two. Make two the denominator of the expression.

$$(3xy)^{\frac{3}{2}}$$

Answer: $\sqrt{(3xy)^3} = (3xy)^{\frac{3}{2}}$

RULE 8: Exponential expressions with a fraction as the exponent can be changed into radical expressions. The numerator of the exponent is the power of the radicand. The denominator of the exponent is the index of the radical expression.

To change an exponential expression with a fractional exponent to a radical expression, follow these three painless steps.

Step 1: Write the base of the exponential expression under a radical sign.

Step 2: Raise the number under the radical sign to the power of the numerator.

Step 3: Make the denominator of the expression the index of the radical.

Change $x^{\frac{1}{3}}$ to a radical expression.

Step 1: Write the base of the exponential expression under a radical sign.
Write x under the radical sign.

$$\sqrt{x}$$

Step 2: Raise the number under the radical sign to the power of the numerator.
Raise x to the first power. Since one is the exponent, you don't have to write it down.

$$\sqrt{x^1} = \sqrt{x}$$

Step 3: Make the denominator of the exponential expression the index of the radical.

Answer: $x^{\frac{1}{3}} = \sqrt[3]{x}$

Change $(3xy)^{\frac{3}{2}}$ to a radical expression.

Step 1: Write the base of the exponential expression under a radical sign.

The quantity $3xy$ is raised to the $\frac{3}{2}$ power, so $3xy$ is the base of the expression.

$$\sqrt{3xy}$$

Step 2: Raise the number under the radical sign to the power of the numerator.

The numerator is 3. Raise $3xy$ to the third power. Be careful—$\sqrt{3xy^3}$ is not the same as $\sqrt{(3xy)^3}$.

$$\sqrt{(3xy)^3}$$

Step 3: Make the denominator of the expression the index of the radical.

The denominator is two. Make two the index of the radical.

$$\sqrt[2]{(3xy)^3}$$

When two is the index you don't need to write it.

$$\sqrt[2]{(3xy)^3} = \sqrt{(3xy)^3}$$

Answer: $(3xy)^{\frac{3}{2}} = \sqrt{(3xy)^3}$

BRAIN TICKLERS
Set # 41

Change the radical expressions to exponential expressions.

1. $\sqrt{5}$

2. $\sqrt{x}$

3. $\sqrt[3]{12}$

4. $\sqrt[5]{7^2}$

5. $\sqrt[3]{(2x)^2}$

(Answers are on page 203.)

BRAIN TICKLERS
Set # 42

Change the exponential expressions to radical expressions.

1. $10^{\frac{1}{2}}$

2. $3^{\frac{1}{3}}$

3. $(5xy)^{\frac{1}{2}}$

4. $7^{\frac{2}{3}}$

5. $(9x)^{\frac{2}{3}}$

(Answers are on page 204.)

SUPER BRAIN TICKLERS

Solve.

1. $\sqrt[3]{125}$

2. $\sqrt[3]{-27}$

3. $\sqrt{12}$

4. $\sqrt{-49}$

5. $\sqrt{(8)(2)}$

6. $\sqrt{12}\,\sqrt{3}$

7. $\sqrt{32}$

8. $\frac{8}{\sqrt{5}}$

9. $2\sqrt{3} - 4\sqrt{3}$

10. $(5\sqrt{2})(2\sqrt{18})$

(Answers are on page 204.)

WORD PROBLEMS

Watch as the following word problems that use roots and radicals are solved.

PROBLEM 1: The square root of a number plus two times the square root of the same number is twelve. What is the number?

First change this problem from Plain English to Math Talk.

"The square root of a number" becomes "$\sqrt{x}$."

"Plus" becomes "+."

"Two times the square root of the same number" becomes "$2\sqrt{x}$."

"Is" becomes "=."

"Twelve" becomes "12."

Now you can change this problem into an equation.
$\sqrt{x} + 2\sqrt{x} = 12$

Now solve this equation.

Add $\sqrt{x} + 2\sqrt{x}$. You can add these two expressions because they have the same radicand and the same index.

$\sqrt{x} + 2\sqrt{x} = 3\sqrt{x}$

The new equation is $3\sqrt{x} = 12$.

Divide both sides of the equation by 3.

$\dfrac{3\sqrt{x}}{3} = \dfrac{12}{3}$

Compute.

$\sqrt{x} = 4$

Square both sides of the equation.

$(\sqrt{x})^2 = (4)^2$

Compute.

$x = 16$

The number is 16.

PROBLEM 2: Seth lives in Bethesda. Sarah lives nine miles east of Seth. Tara lives twelve miles north of Sarah. How far does Seth live from Tara?

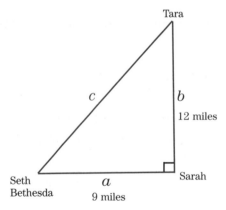

The first step is to change this problem from Plain English to Math Talk. To do this, make a drawing of the problem. The drawing should be a right triangle. The distance from Seth's house to Tara's house is the hypotenuse of the right triangle.

To find this distance use the formula $a^2 + b^2 = c^2$.

"a" is the distance from Seth's house to Sarah's house, which is 9.

"b" is the distance from Sarah's house to Tara's house, which is 12.

"c" is the distance from Seth's house to Tara's house, which is unknown.

Substitute these values in the equation to solve the problem.
$9^2 + (12)^2 = c^2$
Square both the nine and the twelve.
$9^2 = 81$ and $(12)^2 = 144$

Substitute these values into the original equation.
$81 + 144 = c^2$
$225 = c^2$

Take the square root of both sides of this equation.
$15 = c$
It is 15 miles from Seth's house to Tara's house.

BRAIN TICKLERS—
THE ANSWERS

Set # 32, page 168

1. $\sqrt{9} = 3$

2. $\sqrt{25} = 5$

3. $\sqrt{36} = 6$

4. $\sqrt{81} = 9$

5. $\sqrt{49} = 7$

6. $\sqrt{100} = 10$

7. $\sqrt{16} = 4$

8. $\sqrt{4} = 2$

9. $\sqrt{64} = 8$

10. $\sqrt{1} = 1$

Set # 33, page 172

1. $\sqrt[3]{27} = 3$

2. $\sqrt[3]{64} = 4$

3. $\sqrt[5]{1} = 1$

4. $\sqrt[4]{16} = 2$

5. $\sqrt[3]{125} = 5$

6. $\sqrt[10]{0} = 0$

Set # 34, page 173

1. If $\sqrt[x]{16} = 4$, then $x = 2$.

2. If $\sqrt[x]{16} = 2$, then $x = 4$.

3. If $\sqrt[x]{25} = 5$, then $x = 2$.

4. If $\sqrt[x]{125} = 5$, then $x = 3$.

5. If $\sqrt[x]{8} = 2$, then $x = 3$.

Set # 35, page 177

1. $\sqrt{-64}$ is undefined.

2. $\sqrt[4]{-16}$ is undefined.

3. $\sqrt[3]{-27} = -3$

4. $\sqrt[5]{-32} = -2$

5. $\sqrt{-49}$ is undefined.

6. $\sqrt[9]{-1} = -1$

Set # 36, page 180

1. $\sqrt{(16)(9)} = 12$

2. $\sqrt{(64)(100)} = 80$

3. $\sqrt{25y^2} = 5y$

4. $\sqrt{(4)(11)} = 2\sqrt{11}$

5. $\sqrt{9y} = 3\sqrt{y}$

Set # 37, page 182

1. $(\sqrt{3})(\sqrt{3}) = 3$

2. $(\sqrt{8})(\sqrt{2}) = 4$

3. $(\sqrt{12})(\sqrt{3}) = 6$

4. $(\sqrt{x})(\sqrt{x}) = x$

5. $(\sqrt{x^3})(\sqrt{x}) = x^2$

Yep. You're really cookin'!

Set # 38, page 184

1. $\sqrt{20} = 2\sqrt{5}$

2. $\sqrt{8} = 2\sqrt{2}$

3. $\sqrt{27} = 3\sqrt{3}$

4. $\sqrt{24} = 2\sqrt{6}$

5. $\sqrt{32} = 4\sqrt{2}$

6. $\sqrt{125} = 5\sqrt{5}$

Set # 39, page 188

1. $\frac{5}{\sqrt{3}} = \frac{5\sqrt{3}}{3}$

2. $\frac{10}{\sqrt{2}} = \frac{10\sqrt{2}}{2} = 5\sqrt{2}$

3. $\sqrt{\frac{16}{3}} = \frac{4\sqrt{3}}{3}$

4. $\sqrt{\frac{36}{7}} = \frac{6\sqrt{7}}{7}$

Set # 40, page 191

1. $7\sqrt{2} + \sqrt{2} = 8\sqrt{2}$

2. $3\sqrt{3} + \sqrt{3} = 4\sqrt{3}$

3. $5\sqrt[4]{x} + 2\sqrt[4]{x} = 7\sqrt[4]{x}$

4. $9\sqrt{5} - \sqrt{5} = 8\sqrt{5}$

5. $2\sqrt{x} - 2\sqrt{x} = 0$

6. $5\sqrt{2x} - 2\sqrt{2x} = 3\sqrt{2x}$

Set # 41, page 196

1. $\sqrt{5} = 5^{\frac{1}{2}}$

2. $\sqrt{x} = x^{\frac{1}{2}}$

3. $\sqrt[3]{12} = 12^{\frac{1}{3}}$

4. $\sqrt[5]{7^2} = 7^{\frac{2}{5}}$

5. $\sqrt[3]{(2x)^2} = (2x)^{\frac{2}{3}}$

Set # 42, page 196

1. $10^{\frac{1}{2}} = \sqrt{10}$

2. $3^{\frac{1}{3}} = \sqrt[3]{3}$

3. $(5xy)^{\frac{1}{2}} = \sqrt{5xy}$

4. $7^{\frac{2}{3}} = \sqrt[3]{7^2}$

5. $(9x)^{\frac{2}{3}} = \sqrt[3]{(9x)^2}$

Super Brain Ticklers, page 197

1. $\sqrt[3]{125} = 5$

2. $\sqrt[3]{-27} = -3$

3. $\sqrt{12} = 2\sqrt{3}$

4. $\sqrt{-49}$ is undefined.

5. $\sqrt{(8)(2)} = 4$

6. $\sqrt{12}\,\sqrt{3} = 6$

7. $\sqrt{32} = 4\sqrt{2}$

8. $\dfrac{8}{\sqrt{5}} = \dfrac{8\sqrt{5}}{5}$

9. $2\sqrt{3} - 4\sqrt{3} = -2\sqrt{3}$

10. $(5\sqrt{2})(2\sqrt{18}) = 60$

CHAPTER SEVEN

Quadratic Equations

A quadratic equation is an equation with a variable to the second power but no variable higher than the second power. A quadratic equation has the form $ax^2 + bx + c = 0$, where a is not equal to zero. Here are five examples of quadratic equations. In all of these quadratic equations, there is an x^2 term.

$$x^2 + 3x + 5 = 0$$
$$3x^2 - 4x + 3 = 0$$
$$-5x^2 - 2x = 7$$
$$x^2 + 3x = 0$$

(Notice that this equation does not have a numerical term.)

$$x^2 - 36 = 0$$

(Notice that this equation does not have an x term.)

The following are not quadratic equations.

$x^2 - 4x - 6$ is not a quadratic equation because it does not have an equals sign.

$x^3 - 4x^2 + 2x - 1 = 0$ is not a quadratic equation because there is an x^3 term.

$2x - 6 = 0$ is not a quadratic equation because there is no x^2 term.

Quadratic equations are formed when a linear equation is multiplied by the variable in the equation.

Multiply the linear equation $x + 3 = 0$ by x, and the result is the quadratic equation $x^2 + 3x = 0$.

Multiply $y - 7 = 0$ by $2y$ and the result is $2y^2 - 14y = 0$.

Multiply $x - 6 = 0$ by x and the result is $x^2 - 6x = 0$.

Have you got this?

BRAIN TICKLERS
Set # 43

Multiply the following expressions together to form quadratic equations.

1. $x(3x + 1) = 0$

2. $2x(x - 5) = 5$

3. $-x(2x - 6) = 0$

4. $4x(2x - 3) = -3$

(Answers are on page 248.)

Quadratic equations are also formed when two binomial expressions of the form $x - a$ or $x + a$ are multiplied together. The following binomial expressions are quadratic equations.

$$(x - 3)(x - 2) = 0$$
$$(x + 6)(x + 5) = 0$$
$$(2x - 5)(3x + 4) = 0$$

To multiply two binomial terms together, complete the following five steps.

Step 1: Multiply the two **First** terms together.

Step 2: Multiply the two **Outside** terms together.

Step 3: Multiply the two **Inside** terms together.

Step 4: Multiply the two **Last** terms together.

Step 5: Add the terms together and simplify.

Watch how these two expressions are multiplied together.

$(x - 3)(x + 2) = 0$

Step 1: Multiply the two **First** terms together.
$$(\boldsymbol{x} - 3)(\boldsymbol{x} + 2) = 0$$
The two first terms are x and x.
$$(x)(x) = x^2$$

Step 2: Multiply the two **Outside** terms together.
$$(\boldsymbol{x} - 3)(x + \boldsymbol{2}) = 0$$
The two outside terms are x and 2.
$$(2)(x) = 2x$$

Step 3: Multiply the two **Inside** terms together.
$$(x - \boldsymbol{3})(\boldsymbol{x} + 2) = 0$$
The two inside terms are -3 and x.
$$(-3)(x) = -3x$$

Step 4: Multiply the two **Last** terms together.
$$(x - \boldsymbol{3})(x + \boldsymbol{2}) = 0$$
The last two terms are (2) and (-3).
$$(2)(-3) = -6$$

Step 5: Add the terms together and simplify.
First, the terms are added together.
$$x^2 + 2x - 3x - 6 = 0$$
Next, the expression is simplified.
$$x^2 - x - 6 = 0$$

Watch how these two binomial expressions are multiplied together.

$$(x - 5)(x + 5) = 0$$

Step 1: Multiply the two **First** terms together.
$$(x - 5)(x + 5) = 0$$
The first terms of this expression are x and x.
$$(x)(x) = x^2$$

Step 2: Multiply the two **Outside** terms together.
$$(x - 5)(x + 5) = 0$$
The two outside terms are x and 5.
$$(5)(x) = 5x$$

Step 3: Multiply the two **Inside** terms together.
$$(x - 5)(x + 5) = 0$$
The two inside terms are -5 and x.
$$(-5)(x) = -5x$$

Step 4: Multiply the two **Last** terms together.
$$(x - 5)(x + 5) = 0$$
The last two terms are -5 and 5.
$$(-5)(5) = -25$$

Step 5: Add the terms together and simplify.
First, add the terms together.
$$x^2 + 5x - 5x - 25 = 0$$
Simplify this expression.
$$x^2 - 25 = 0$$

There it is!
The LAST clue!

Watch how these two binomial expressions are multiplied together.

$$(3x - 5)(2x - 7) = 0$$

Step 1: Multiply the two **First** terms together.
$$(\mathbf{3x} - 5)(\mathbf{2x} - 7) = 0$$
The two first terms are $3x$ and $2x$.
$$(3x)(2x) = 6x^2$$

Step 2: Multiply the two **Outside** terms together.
$$(\mathbf{3x} - 5)(2x - \mathbf{7}) = 0$$
The two outside terms are $3x$ and -7.
$$(3x)(-7) = -21x$$

Step 3: Multiply the two **Inside** terms together.
$$(3x - \mathbf{5})(\mathbf{2x} - 7) = 0$$
The two inside terms are -5 and $2x$.
$$(-5)(2x) = -10x$$

Step 4: Multiply the two **Last** terms together.
$$(3x - \mathbf{5})(2x - \mathbf{7}) = 0$$
The last two terms are -5 and -7.
$$(-5)(-7) = 35$$

Step 5: Add the terms together and simplify.
Add the terms together.
$$6x^2 - 21x - 10x + 35 = 0$$
Simplify.
$$6x^2 - 31x + 35 = 0$$

SUM IT UP!

Remember the steps in multiplying two binomial expressions. Multiply the parts of two binomial expressions together in the following order.

FIRST, OUTSIDE, INSIDE, LAST

Remember the order by remembering the word **FOIL**. The word **FOIL** is the first letters of the words, **F**irst, **O**utside, **I**nside, **L**ast.

BRAIN TICKLERS
Set # 44

Multiply these binomial expressions together to form quadratic equations.

1. $(x + 5)(x + 2) = 0$

2. $(x - 3)(x + 1) = 0$

3. $(2x - 5)(3x + 1) = 0$

4. $(x + 2)(x - 2) = 0$

(Answers are on page 248.)

SOLVING QUADRATIC EQUATIONS BY FACTORING

Now that you know how to recognize quadratic equations, how do you solve them? The easiest way to solve most quadratic equations is by factoring. Before factoring a quadratic equation, you must put it in standard form. Standard form is $ax^2 + bx + c = 0$, where a, b, and c can be any real numbers, except that a cannot equal zero.

Here are four examples of quadratic equations in standard form.

$$x^2 + x - 1 = 0$$
$$5x^2 + 3x - 2 = 0$$
$$-2x^2 - 2 = 0 \quad \text{In this equation, } b = 0.$$
$$4x^2 - 2x = 0 \quad \text{In this equation, } c = 0.$$

Here are three examples of quadratic equations *not* in standard form.

$$3x^2 - 7x = 4$$
$$x^2 = 3x + 2$$
$$x^2 = 4$$

If an equation is not in standard form, you can put it in standard form by adding and/or subtracting the same terms from both sides of the equation.

Watch as these quadratic equations are put in standard form.

$3x^2 - 5x = 2$
To put this equation in standard form, subtract 2 from both sides of the equation.
$$3x^2 - 5x - 2 = 2 - 2$$
Simplify.
$$3x^2 - 5x - 2 = 0$$

$x^2 = 2x - 1$
To put this equation in standard form, subtract $2x$ from both sides of the equation.
$$x^2 - 2x = 2x - 2x - 1$$
Simplify.
$$x^2 - 2x = -1$$
Now add 1 to both sides of the equation.
$$x^2 - 2x + 1 = -1 + 1$$
Simplify the equation.
$$x^2 - 2x + 1 = 0$$

$$4x^2 = 2x$$

To put this equation in standard form, subtract $2x$ from both sides of the equation.

$$4x^2 - 2x = 2x - 2x$$

Simplify the equation.

$$4x^2 - 2x = 0$$

BRAIN TICKLERS
Set # 45

Change the following quadratic equations to standard form.

1. $x^2 + 4x = -6$

2. $2x^2 = 3x - 3$

3. $5x^2 = -5x$

4. $7x^2 = 7$

(Answers are on page 248.)

Once the equation is in standard form, you can solve it by factoring. Remember, in standard form a quadratic equation has the form $ax^2 + bx + c = 0$, where a is not equal to zero. There are three types of quadratic equations in standard form. You will learn how to solve each of them separately.

Type I: A Type I quadratic equation has only two terms. It has the form $ax^2 + c = 0$. A Type I quadratic equation has no middle term, so $b = 0$.

Type II: A Type II quadratic equation has only two terms. It has the form $ax^2 + bx = 0$. A Type II quadratic equation has no last term, so $c = 0$.

Type III: A Type III quadratic equation has all three terms. It has the form $ax^2 + bx + c = 0$. In a quadratic equation of this type, a is not equal to zero, b is not equal to zero, and c is not equal to 0.

Now let's see how to solve each of these types.

Type I: In Type I quadratic equations, b is equal to zero. When $b = 0$, the quadratic equation has no x term. Here are examples of quadratic equations where $b = 0$.

$$x^2 - 36 = 0$$
$$x^2 - 25 = 0$$
$$x^2 + 2 = 0$$
$$2x^2 - 18 = 0$$

Remember!
Type I has no
middle term.

To solve Type I equations, use the following four steps.

Step 1: Add or subtract to move the x^2 term to one side of the equals sign and the numerical term to the other side of the equals sign.

Step 2: Multiply or divide to eliminate the coefficient in front of the x^2 term.

Step 3: Take the square root of both sides of the equals sign. When solving quadratic equations, the symbol $+/-$ means use *both* the positive and the negative square roots of the number.

Step 4: Check your answer.

Watch how the following quadratic equation is solved.

Solve $x^2 - 36 = 0$.

Step 1: Add or subtract to move the x^2 term on one side of the equals sign and the numerical term on the other side of the equals sign.
Add 36 to both sides of the equation.
$$x^2 - 36 + 36 = 0 + 36$$
Simplify:
$$x^2 = 36$$

Step 2: Multiply or divide to eliminate the coefficient in front of the x^2 term.
There is no coefficient in front of the x^2 term. Go to the next step.

Step 3: Take the square root of both sides of the equation.
$$\sqrt{x^2} = \sqrt{36}$$
The square root of x squared is x. The square roots of 36 are 6 and -6.
Solution: If $x^2 = 36$, then $x = 6$ or $x = -6$.

Step 4: Check your answer.

Substitute 6 for x in the original equation.
The original equation is $x^2 - 36 = 0$
Put a 6 wherever there is an x.
$$(6)^2 - 36 = 0$$
Simplify.
$$36 - 36 = 0$$
$$0 = 0 \text{ is a true sentence.}$$
So, $x = 6$ is a solution to the equation $x^2 - 36 = 0$.

Substitute -6 for x in the original equation.
The original equation is $x^2 - 36 = 0$
Put a -6 wherever there is an x.
$$(-6)^2 - 36 = 0$$
Simplify.
$$36 - 36 = 0$$
$$0 = 0 \text{ is a true sentence.}$$
So, $x = -6$ is a solution to the equation $x^2 - 36 = 0$.

Watch as another example is solved.

Solve $2x^2 - 18 = 0$.

Step 1: Add or subtract to move the x^2 term on one side of the equals sign and the numerical term on the other side of the equals sign. Add 18 to both sides of the equation.
$$2x^2 - 18 + 18 = 0 + 18$$
Simplify.
$$2x^2 = 18$$

Step 2: Multiply or divide to eliminate the coefficient in front of the x^2 term.
Divide both sides of the equation by two.
$$\frac{2x^2}{2} = \frac{18}{2}$$
Simplify.
$$x^2 = 9$$

Step 3: Take the square root of both sides of the equation.
$$\sqrt{x^2} = \sqrt{9}$$
The square root of x^2 is x. The square roots of 9 are 3 and -3.
Solution: If $2x^2 = 18$, then $x = 3$ or $x = -3$.

Step 4: Check your answer.
Substitute 3 for x back in the original equation.
The original equation is $2x^2 - 18 = 0$.
Put 3 wherever there is an x.
$$2(3)^2 - 18 = 0$$
$$2(9) - 18 = 0$$
$$18 - 18 = 0 \text{ is a true sentence.}$$
So, $x = 3$ is a solution for the equation $2x^2 - 18 = 0$.

Now substitute -3 for x back in the original equation.
The original equation is $2x^2 - 18 = 0$.
Put -3 wherever there is an x.
$$2(-3)^2 - 18 = 0$$
$$2(9) - 18 = 0$$
$$18 - 18 = 0 \text{ is a true sentence.}$$
So, $x = -3$ is a solution for the equation $2x^2 - 18 = 0$.

Watch as one more problem is solved.

Solve $4x^2 - 8 = 0$.

Step 1: Add or subtract to put the x^2 term on one side of the equals sign and the numerical term on the other side of the equals sign.
Add 8 to both sides of the equation.
$$4x^2 - 8 + 8 = 0 + 8$$
Simplify.
$$4x^2 = 8$$

Step 2: Multiply or divide to eliminate the coefficient in front of the x^2 term.

Divide both sides of the equation by 4.

$$\frac{4x^2}{4} = \frac{8}{4}$$

Simplify.

$$x^2 = 2$$

Step 3: Take the square root of both sides of equation.

$$\sqrt{x^2} = \sqrt{2}$$

The square root of x^2 is x. Two is not a perfect square, so the 2 is $+2$ and -2.

Solution: If $4x^2 = 8$, then $x = \sqrt{2}$ or $x = -\sqrt{2}$.

Step 4: Check your answer.

Substitute $\sqrt{2}$ for x back in the original equation.

The original equation is $4x^2 - 8 = 0$.

Put $\sqrt{2}$ wherever there is an x.

$$4(\sqrt{2})^2 - 8 = 0$$
$$4(2) - 8 = 0$$
$$8 - 8 = 0 \text{ is a true sentence.}$$

So, $x = 2$ is a solution for the equation $4x^2 - 8 = 0$.

Substitute $-\sqrt{2}$ for x back in the original equation.

The original equation is $4x^2 - 8 = 0$.

Put $-\sqrt{2}$ wherever there is an x.

$$4(-\sqrt{2})^2 - 8 = 0$$
$$4(2) - 8 = 0$$
$$8 - 8 = 0 \text{ is a true sentence.}$$

So, $x = -2$ is also a solution for the equation $4x^2 - 8 = 0$.

BRAIN TICKLERS
Set # 46

Solve the following Type I Quadratic Equations.

1. $x^2 - 25 = 0$ 4. $2x^2 - 32 = 0$

2. $x^2 - 49 = 0$ 5. $x^2 - 15 = 0$

3. $3x^2 - 27 = 0$ 6. $3x^2 - 20 = 10$

(Answers are on page 248.)

Type II: In Type II quadratic equations, c is equal to zero. When c is equal to zero, the quadratic equation has no numerical term. Here are examples of quadratic equations where c is equal to zero.
$$x^2 + 10x = 0$$
$$x^2 + 5x = 0$$
$$x^2 - 3x = 0$$

Remember, Type II has no last term.

To solve Type II equations, follow the following four steps.

Step 1: Factor an x out of the equation.

Step 2: Set both factors equal to zero.

Step 3: Solve both equations.

Step 4: Check your answer.

Watch while the following Type II quadratic equations are solved.

Solve $x^2 - 5x = 0$.

Step 1: Factor an x out of the equation.
$$x(x - 5) = 0$$

Step 2: Set both factors equal to zero.
$$x = 0; x - 5 = 0$$

Step 3: Solve both equations.
The equation $x = 0$ is solved.
To solve $x - 5 = 0$, add 5 to both sides of the equation.
$$x - 5 + 5 = 0 + 5$$
Simplify.
$$x = 5$$
Solution: If $x^2 - 5x = 0$, then $x = 5$ or $x = 0$.

Step 4: Check your answer.
Substitute 5 for x in the original equation.
The original equation is $x^2 - 5x = 0$
$$(5)^2 - 5(5) = 0$$
$$25 - 25 = 0$$
$$0 = 0 \text{ is a true sentence.}$$
So, $x = 5$ is a correct answer.

Now substitute 0 for x in the original equation. The original equation is $x^2 - 5x = 0$.
$$0^2 - 5(0) = 0$$
$0 = 0$ is a true sentence.
So, $x = 5$ and $x = 0$ are both solutions for the equation.

Watch closely as another example is solved.

Solve $2x^2 - 12x - 0$.

Step 1: Factor x out of the equation.
$$x(2x - 12) = 0$$

Step 2: Set both factors equal to zero.
$$x = 0; 2x - 12 = 0$$

Step 3: Solve both equations.
$x = 0$ is already solved.
Solve $2x - 12 = 0$. Add 12 to both sides of this equation.
$$2x - 12 + 12 = 0 + 12$$
Simplify.
$$2x = 12$$
Divide both sides by 2.
$$x = 6$$

Step 4: Check the answers.
Substitute 6 for x in the original equation, $2x^2 - 12x = 0$.
$$2(6)^2 - 12(6) = 0$$
$$2(36) - 12(6) = 0$$
$$72 - 72 = 0$$
$$0 = 0$$
Substitute 0 for x in the original equation, $2x^2 - 12x = 0$.
$$2(0)^2 - 12(0) = 0$$
$$0 = 0$$
So, $x = 6$ and $x = 0$ are both solutions to the original equation.

BRAIN TICKLERS
Set # 47

Solve the following equations.

1. $x^2 - 2x = 0$

2. $x^2 + 4x = 0$

3. $2x^2 - 6x = 0$

4. $\frac{1}{2}x^2 + 2x = 0$

(Answers are on page 249.)

Type III: In a quadratic equation of this type, a is not equal to zero, b is not equal to zero, and c is not equal to zero. In standard form, a quadratic equation of this type has three terms. Here are examples of quadratic equations of this type.

$$x^2 + 5x + 6 = 0$$
$$x^2 - 2x + 1 = 0$$
$$x^2 - 3x - 4 = 0$$

In order to solve a quadratic equation with three terms, factor it into two binomial expressions. Follow these steps.

Step 1: Put the equation in standard form.

Step 2: Factor the x^2 term.

Step 3: List the pairs of factors of the numerical term.

Step 4: Place one pair of factors in parentheses. Check to see if when the two binomial expressions are multiplied together, the result is the original equation. If not, try another pair of factors until the correct pair is found.

Step 5: Set each of the two binomial expressions equal to zero.

Step 6: Solve each of the equations.

Step 7: Check your answers.

Watch while this quadratic equation with all three terms is solved.

Solve $x^2 + 2x + 1 = 0$.

Step 1: Put the equation in standard form.
This equation is in standard form.
Go on to the next step.

Step 2: Draw two sets of parentheses and factor the x^2 term. Place the factors in the parentheses.
$$x^2 = (x)(x)$$
$$(x \quad)(x \quad) = 0$$

Step 3: List the possible pairs of factors of the numerical term.
The numerical term is 1.
What are the possible factors of 1?
$$(1)(1) = 1 \text{ and } (-1)(-1) = 1$$

Step 4: Place one pair of factors in the parentheses. Check to see if when the two binomial expressions are multiplied together, the result is the original equation. If the result is the original equation, go to Step 5. If the result is not the original equation, try another pair of factors.

Remember, Type III has all three terms and they're not equal to zero.

Place (1) and (1) into the parentheses.
$$(x + 1)(x + 1) = 0$$
Multiply these binomial expressions together. Multiply the first terms, outside terms, inside terms, and last terms together (FOIL). Add the results together.
$$(x + 1)(x + 1) = x^2 + 1x + 1x + 1 = 0$$
Simplify.
$$x^2 + 2x + 1 = 0$$
This is the original equation.
Therefore, 1 and 1 are the correct factors.

Step 5: Set each of the two binomial expressions equal to zero.
$$x + 1 = 0 \text{ and } x + 1 = 0$$

Step 6: Solve each of the equations.
Both equations are the same, so you only need to solve one of them. Subtract (-1) from both sides of the equation.
$$x + 1 - 1 = 0 - 1$$
Simplify.
$$x = -1$$

Step 7: Check your answer.
Substitute -1 into the original equation.
$$x^2 + 2x + 1 = 0$$
$$(-1)^2 + 2(-1) + (1) = 0$$
$$1 + (-2) + 1 = 0$$
$$0 = 0$$
So, $x = -1$ is the correct answer.

Watch how another equation is solved by factoring.

Solve $x^2 - 5x + 4 = 0$.

Step 1: Put the equation in standard form.
The equation is in standard form.
Go to the next step.

Step 2: Factor the x^2 term and place the factors in parentheses.
$$(x \quad)(x \quad) = 0$$

Step 3: List all the factors for the numerical term, 4.
$$(2)(2) = 4$$
$$(-2)(-2) = 4$$
$$(4)(1) = 4$$
$$(-4)(-1) = 4$$

Step 4: Check to see which pair of factors found in Step 3, when placed in the parentheses and multiplied together, will result in the original equation. Only one pair will work.
Place the factors 2 and 2 in the parentheses.
$$(x + 2)(x + 2) = 0$$
Multiply the two expressions together.
$$x^2 + 2x + 2x + 4 = 0$$
Simplify.
$$x^2 + 4x + 4 = 0$$
This is not the original equation, so try another pair of factors.

Place the factors 4 and 1 in the parentheses.
$$(x + 4)(x + 1) = 0$$
Multiply the two binomial expressions together.
$$x^2 + 1x + 4x + 4 = 0$$
Simplify.
$$x^2 + 5x + 4 = 0$$
This is not the original equation, so try another pair of factors.

Place the factors -2 and -2 inside the parentheses.
$$(x - 2)(x - 2) = 0$$
Multiply the two expressions together.
$$x^2 - 2x - 2x + 4 = 0$$
Simplify.
$$x^2 - 4x + 4 = 0$$
This is not the original equation, so try another pair of factors.

Place the factors -4 and -1 inside the parentheses.
$$(x - 4)(x - 1) = 0$$
Multiply the two expressions together.
$$x^2 - 4x - 1x + 5 = 0$$
Simplify.
$$x^2 - 5x + 5 = 0$$
This is the original equation.
Therefore, -4 and -1 are the correct factors.

Step 5: Set each of the two binomial expressions equal to zero.
$$x - 4 = 0 \text{ and } x - 1 = 0$$

Step 6: Solve each of the equations.
Solve $x - 4 = 0$. Add 4 to both sides of the equation.
$$x - 4 + 4 = 0 + 4$$
$$x = 4$$
Now solve $x - 1 = 0$. Add 1 to both sides of the equation.
$$x - 1 + 1 = 0 + 1$$
$$x = 1$$

Step 7: Check your answers.

Substitute 4 in the original equation, $x^2 - 5x + 4 = 0$.
$$4^2 - 5(4) + 4 = 0$$
Compute the value of this expression.
$$16 - 20 + 4 = 0$$
$$0 = 0$$
This proves that $x = 4$ is a correct solution to the equation.

Now substitute $x = 1$ into the original equation, $x^2 - 5x + 4 = 0$.
$$(1)^2 - 5(1) + 4 = 0$$
Compute the value of this expression.
$$1 - 5 + 4 = 0$$
$$0 = 0$$
So, $x = 1$ is also a solution to the equation $x^2 - 5x + 4 = 0$.
This equation has two solutions, $x = 4$ and $x = 1$.

Now watch as one more quadratic equation is solved by factoring.

Solve $2x^2 + 7x = -6$.

This is still Type III!

Step 1: Put the equation in standard form.
Add six to both sides of the equation.
$$2x^2 + 7x + 6 = -6 + 6$$
Simplify.
$$2x^2 + 7x + 6 = 0$$

Step 2: Factor the x^2 term and place the factors in parentheses. Notice that the x^2 term has a 2 in front of it. The only way to factor $2x^2$ is $(2x)(x)$. Put these terms in the parentheses.
$$(2x \quad)(x \quad) = 0$$

Step 3: List all the factors of the numerical term.
The factors of 6 are
$$(6)(1) = 6$$
$$(-6)(-1) = 6$$
$$(3)(2) = 6$$
$$(-3)(-2) = 6$$

Step 4: Substitute each pair of numerical factors into the parentheses. Check to see if, when the two binomial expressions are multiplied together, the result is the original equation.
Substitute 6 and 1 into the parentheses.
$$(2x + 6)(x + 1) = 0$$
Multiply to see if you get the original equation.
$$2x^2 + 2x + 6x + 6 = 0$$
Simplify.
$$2x^2 + 8x + 6 = 0$$
This is not the original equation, but before testing the next set of factors, SWITCH the position of the numbers.
$$(2x + 1)(x + 6) = 0$$
Multiply the equation.
$$2x^2 + 12x + 1x + 6 = 0$$
Simplify.
$$2x^2 + 13x + 6 = 0$$
By reversing the position of the 6 and the 1, a new equation was formed. But this is not the original equation, so put the next pair of factors into the equation.

Substitute -6 and -1 into the parentheses.
$$(2x - 6)(x - 1) = 0$$
Multiply to see if you get the original equation.
$$2x^2 - 2x - 6x + 6 = 0$$
Simplify:
$$2x^2 - 8x + 6 = 0$$
This is not the original equation, but before testing the next set of factors, SWITCH the position of the numbers.
$$(2x - 1)(x - 6) = 0$$

Multiply the two binomial expressions together.
$$2x^2 - 12x - 1x + 6 = 0$$
Simplify.
$$2x^2 - 13x + 6 = 0$$
By reversing the position of the -6 and -1, a new equation was formed. But this is not the original equation. Put the next pair of factors into the equation.

Substitute 3 and 2 into the parentheses.
$$(2x + 3)(x + 2) = 0$$
Multiply to see if you get the original equation.
$$2x^2 + 4x + 3x + 6 = 0$$
Simplify.
$$2x^2 + 7x + 6 = 0$$
This is the original equation.

Step 5: Set each of the two binomial expressions equal to zero.
$$2x + 3 = 0 \text{ and } x + 2 = 0$$

Step 6: Solve each of the equations.
$$2x + 3 = 0$$
Subtract 3 from both sides of the equation.
$$2x + 3 - 3 = 0 - 3$$
$$2x = -3$$
Divide both sides by 2.
$$\frac{2x}{2} = -\frac{3}{2}$$
$$x = -\frac{3}{2}$$
If $2x + 3 = 0$, then $x = -\frac{3}{2}$.

Solve $x + 2 = 0$.
Subtract 2 from both sides of the equation.
$$x + 2 - 2 = 0 - 2$$
$$x = -2$$
If $x + 2 = 0$, then $x = -2$.

Step 7: Check your answer.

Substitute $x = -\frac{3}{2}$ and $x = -2$ into the original equation.

The original equation is $2x^2 + 7x + 6 = 0$

First substitute $x = -\frac{3}{2}$ in $2x^2 + 7x + 6$.

$$2\left(-\frac{3}{2}\right)^2 + 7\left(-\frac{3}{2}\right) + 6 = 0$$

Square $-\frac{3}{2}$.

$$2\left(\frac{9}{4}\right) + 7\left(-\frac{3}{2}\right) + 6 = 0$$

Multiply.

$$\frac{9}{2} - \frac{21}{2} + 6 = 0$$

Change 6 to $\frac{12}{2}$ and add.

$$\frac{9}{2} - \frac{21}{2} + \frac{12}{2} = 0$$
$$0 = 0$$

So, $x = -\frac{3}{2}$ is a solution to the equation $2x^2 + 7x + 6 = 0$.

Now substitute $x = -2$ into the original equation, $2x^2 + 7x + 6 = 0$.

$$2(-2)^2 + 7(-2) + 6 = 0$$

Square (-2).

$$2(4) + 7(-2) + 6 = 0$$
$$8 + (-14) + 6 = 0$$
$$0 = 0$$

So, $x = -2$ is a solution to the equation $2x^2 + 7x + 6 = 0$.

BRAIN TICKLERS
Set # 48

Solve the following equations by factoring.

1. $x^2 + 10x + 24 = 0$

2. $x^2 + x - 12 = 0$

3. $2x^2 - 7x + 5 = 0$

4. $x^2 - 2x - 3 = 0$

(Answers are on page 249.)

Here is one last example to try.

Solve $x^2 + x = -1$.

Step 1: Put the equation in standard form.
Add 1 to both sides of the equation.
$$x^2 + x + 1 = -1 + 1$$
Simplify.
$$x^2 + x + 1 = 0$$

Step 2: Factor the x^2 term.
There is only one way to factor the x^2 term.
$$(x)(x) = x^2$$
Place these factors in a set of parentheses.
$$(x\ \ )(x\ \ ) = 0$$

Step 3: List the factors of the numerical term.
$$(1)(1) = 1$$
$$(-1)(-1) = 1$$

Step 4: Try substituting each pair of factors into the parentheses. Check to see if, when the two binomial expressions are multiplied together, the result is the original equation.
Substitute 1 and 1 into $(x \quad)(x \quad) = 0$
$$(x + 1)(x + 1) = 0$$
Compute the value of this expression.
$$x^2 + 1x + 1x + 1 = 0$$
Simplify.
$$x^2 + 2x + 1 = 0$$
This is not the original equation.

Try the other set of factors.
Substitute -1 and -1 into the parentheses.
$$(x - 1)(x - 1) = 0$$
Compute the value of this expression.
$$x^2 - 1x - 1x + 1 = 0$$
Simplify this expression.
$$x^2 - 2x + 1 = 0$$
This is not the original equation either.

Some quadratic equations cannot be factored, or they are difficult to factor. In order to solve these equations, the quadratic formula was invented.

THE QUADRATIC FORMULA

Another way to solve a quadratic equation is to use the quadratic formula.

When you use the quadratic formula, you can solve quadratic equations without factoring. Just put the equation in standard form, $ax^2 + bx + c = 0$. Substitute the values for a, b and c into this equation and simplify it.

$$\frac{-b \pm \sqrt{b^2 - 4ac}}{2a}$$

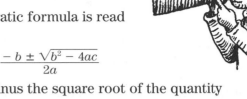

MATH TALK!

Watch how the quadratic formula is read in Plain English.

$$\frac{-b \pm \sqrt{b^2 - 4ac}}{2a}$$

Negative b plus or minus the square root of the quantity b squared minus four times a times c, all divided by two times a.

The result will be the answer. When you solve a quadratic equation with the quadratic formula, you may get zero, one, or two answers. It may look complicated, but it's *painless*. Just follow these four steps.

Step 1: Put the quadratic equation in standard form, $ax^2 + bx + c = 0$.

Step 2: Figure out the values of a, b and c.

In the quadratic equation $2x^2 - 6x + 4 = 0$, $a = 2$, $b = -6$, and $c = 4$.
In the quadratic equation $x^2 - 5x - 3 = 0$, $a = 1$, $b = -5$, and $c = -3$.
In the quadratic equation $4x^2 - 2 = 0$, $a = 4$, $b = 0$, and $c = -2$.
In the quadratic equation $x^2 + 3x = 0$, $a = 1$, $b = 3$, and $c = 0$.

Step 3: Substitute the values of a, b, and c into the quadratic formula to solve for x.

$$\frac{-b \pm \sqrt{b^2 - 4ac}}{2a}$$

Step 4: Check your answer.

Watch as an equation is solved using the quadratic formula.

Solve $x^2 + 2x + 1 = 0$.

Step 1: Put the quadratic equation in standard form, $ax^2 + bx + c = 0$.
The equation $x^2 + 2x + 1 = 0$ is in standard form.
Go to the next step.

Step 2: Figure out the values of a, b and c.

The coefficient in front of the x^2 term is a, so $a = 1$.
The coefficient in front of the x term is b, so $b = 2$.
The number in the equation is c, so $c = 1$.

Step 3: Substitute a, b, and c into the quadratic formula and solve it.
$$\frac{-b \pm \sqrt{b^2 - 4ac}}{2a} = \frac{-2 \pm \sqrt{2^2 - 4(1)(1)}}{2(1)}$$

To compute the value of this expression, compute the value of the numbers under the radical sign first.

The expression under the radical sign is $2^2 - 4(1)(1) = 4 - 4 = 0$.
The value under the radical sign is 0.

Substitute zero for the expression under the radical sign.
$$\frac{-2 \pm 0}{2(1)}$$
Divide.
$$\frac{-2}{2} = -1$$

Step 4: Check your answer.
Substitute -1 into the original equation.
If the result is a true sentence, then -1 is the correct answer.
Substitute -1 into $x^2 + 2x + 1 = 0$.
$$(-1)^2 + 2(-1) + 1 = 0$$
Compute.
$$1 + (-2) + 1 = 0$$
$$0 = 0$$
So, $x = -1$ is the correct answer.
Therefore, $x^2 + 2x + 1 = 0$ when $x = -1$.

Watch as another equation is solved using the quadratic formula.

Solve $2x^2 + 5x + 2 = 0$.

Step 1: Put the equation in standard form.
The equation is in standard form.
Go to the next step.

Step 2: Figure out the values of a, b, and c.
The coefficient in front of the x^2 term is a, so $a = 2$.
The coefficient in front of the x term is b, so $b = 5$.
The number in the equation is c, so $c = 2$.

Step 3: Substitute the values of a, b, and c into the quadratic formula.
$$\frac{-b \pm \sqrt{b^2 - 4ac}}{2a} = \frac{-5 \pm \sqrt{(5)^2 - 4(2)(2)}}{2(2)}$$
Compute the value under the radical.
$$(5)^2 = 25 \text{ and } (4)(2)(2) = 16$$
$$\sqrt{25 - 16} = \sqrt{9} = 3$$
Substitute 3 into the equation.
$$\frac{-5 \pm 3}{4}$$

The numerator of this expression, -5 ± 3, is read as "negative five plus or minus three." This expression makes two separate expressions.
$$\frac{-5 + 3}{4} \text{ and } \frac{-5 - 3}{4}$$

Now the two expressions read
negative five *plus* three, all divided by four;
negative five *minus* three, all divided by four.

Compute the value of these two expressions.
$$\frac{-5 + 3}{4} = \frac{-2}{4} = -\frac{1}{2} \qquad \frac{-5 - 3}{4} = \frac{-8}{4} = -2$$

The two possible solutions are $-\frac{1}{2}$ and -2.

To check if $-\frac{1}{2}$ and/or -2 are the correct answers, substitute them into the original equation. If either result is a true sentence, then the corresponding answer is a correct answer.

Step 4: Check your answer.

Start by substituting $-\frac{1}{2}$ in the original equation.

$$2x^2 + 5x + 2 = 0$$

$$2\left(-\frac{1}{2}\right)^2 + 5\left(-\frac{1}{2}\right) + 2 = 0$$

Compute. Using the order of operations, compute the value of the exponents first.

$$\left(-\frac{1}{2}\right)^2 = \frac{1}{4}$$

Substitute $\frac{1}{4}$ for $\left(-\frac{1}{2}\right)^2$.

$$2\left(\frac{1}{4}\right) + 5\left(-\frac{1}{2}\right) + 2 = 0$$

Now multiply.

$$\frac{2}{4} + \left(-\frac{5}{2}\right) + 2 = 0$$

Because $\frac{2}{4}$ is equal to $\frac{1}{2}$, substitute $\frac{1}{2}$ for $\frac{2}{4}$.

$$\frac{1}{2} + \left(-\frac{5}{2}\right) + 2 = 0$$

Compute.

$$-\frac{4}{2} + 2 = 0$$

$$0 = 0$$

This is a true sentence, so $-\frac{1}{2}$ is a correct solution.

Substitute the other possible answer in the original equation.

Substitute -2 into the original equation, $2x^2 + 5x + 2 = 0$.

$$2(-2)^2 + 5(-2) + 2 = 0$$

Using the order of operations, the first step is clearing the exponent. Because $(-2)^2 = 4$, substitute 4 for the first term.

$$2(4) + 5(-2) + 2 = 0$$

Next multiply the terms.
$$(2)(4) = 8; \ 5(-2) = (-10)$$

Substitute these terms.
$$8 + (-10) + 2 = 0$$
$$0 = 0$$

Therefore -2 is a correct answer.

Sometimes a solution to a quadratic equation is also called the *root*, so $-\frac{1}{2}$ and -2 are the *roots* of $2x^2 + 5x + 2 = 0$.

Let's check the root!

Here is another example using the quadratic formula.

Solve $3x^2 + x - 2 = 0$.

Step 1: Put the equation in standard form.

Step 2: Figure out the values of a, b, and c.
The coefficient in front of the x^2 term is a, so $a = 3$.
The coefficient in front of the x term is b, so $b = 1$.
The number in the equation is c, so $c = -2$.

QUADRATIC EQUATIONS

Step 3: Substitute these values into the quadratic formula.

$$\frac{-b \pm \sqrt{b^2 - 4ac}}{2a} = \frac{-1 \pm \sqrt{1^2 - 4(3)(-2)}}{2(3)}$$

To compute the value of this expression, compute the value of the numbers under the radical sign first.
$1^2 - 4(3)(-2)$ is under the radical sign.

$$1 - (-24) = 25$$

The value under the radical sign is 25.
The square root of 25 is 5.

Substitute 5 for the radical expression.

$$\frac{-1 \pm 5}{2(3)}$$

Separate this expression into two separate expressions. One should have a plus sign and one should have a minus sign. Compute the value of each.

$$\frac{-1 + 5}{6} = \frac{4}{6} = \frac{2}{3} \qquad \frac{-1 - 5}{6} = \frac{-6}{6} = -1$$

The two possible solutions are $x = \frac{2}{3}$ and $x = -1$.

To check if $\frac{2}{3}$ and/or -1 are the correct answers, substitute each of them into the original equation. If either result is a true sentence, then the corresponding answer is a correct solution for the equation $3x^2 + x - 2 = 0$.

Step 4: Check your answers.

Substitute $\frac{2}{3}$ into the original equation, $3x^2 + x - 2 = 0$.

$$3\left(\frac{2}{3}\right)^2 + \left(\frac{2}{3}\right) - 2 = 0$$

Compute the value of the exponential expression first.

$$\left(\frac{2}{3}\right)^2 = \frac{4}{9}$$

Substitute $\frac{4}{9}$ for $\left(\frac{2}{3}\right)^2$.

$$3\left(\frac{4}{9}\right) + \left(\frac{2}{3}\right) - 2 = 0$$

Next, multiply.

$$(3)\left(\frac{4}{9}\right) = \frac{12}{9} = \frac{4}{3}$$

Substitute $\frac{4}{3}$ for $(3)\left(\frac{4}{9}\right)$.

$$\frac{4}{3} + \frac{2}{3} - 2 = 0$$

Next, add and subtract.

$$\frac{4}{3} + \frac{2}{3} = \frac{6}{3} = 2$$
$$2 - 2 = 0$$
$$0 = 0$$

So, $\frac{2}{3}$ is a solution to the equation $3x^2 + x - 2 = 0$.

Check if -1 is a correct answer. Substitute -1 for x in the equation $3x^2 + x - 2 = 0$.
$$3(-1)^2 + (-1) - 2 = 0$$
Compute the value of the exponential expression first.
$$(-1)^2 = 1$$
Substitute (1) for $(-1)^2$.
$$3(1) + (-1) - 2 = 0$$
Multiply $(3)(1)$.
$$(3)(1) = 3$$
Substitute 3 for $(3)(1)$.
$$3 + (-1) - 2 = 0$$
Add and subtract.
$$0 = 0$$
So, -1 is a solution to the equation $3x^2 + x - 2 = 0$.

Caution—Major Mistake Territory!

± is read as "plus or minus." When you see this expression you will get two answers.

$$5 \pm 3 = 5 + 3 \text{ or } 5 - 3$$
$$5 + 3 = 8; \ 5 - 3 = 2$$

The two answers, 5 ± 3, are 8 and 2.

BRAIN TICKLERS
Set # 49

Solve the following quadratic equations using the quadratic formula.

1. $x^2 + 4x + 3 = 0$

2. $x^2 - x - 2 = 0$

3. $x^2 - 3x + 2 = 0$

4. $4x^2 + 4x + 1 = 0$

5. $x^2 = 36$

(Answers are on page 249.)

WORD PROBLEMS

Here are two word problems that require factoring to solve.
Read each one carefully and watch how it is solved. It's
painless.

PROBLEM 1: The product of two consecutive even integers is 48.
Find the integers.

First change this problem from Plain English to Math Talk.
The phrase "the product" means multiply.
Let "x" represent the first even integer.
The second even integer is "$x + 2$."
The word "is" means "$=$."
"48" goes on the other side of the equals sign.
Now the problem can be written as $(x)(x + 2) = 48$

To solve, multiply.
$x^2 + 2x = 48$

Put the equation in standard form.
$x^2 + 2x - 48 = 0$
Factor.
$(x + 8)(x - 6) = 0$

Solve.
If $x + 8 = 0$, then $x = -8$.
If $x - 6 = 0$, then $x = 6$.

The two consecutive even integers were named "x" and
"$x + 2$."
If $x = -8$, then $x + 2 = -6$.
If $x = 6$, then $x + 2 = 8$.

Check both pairs of answers.
$(-8)(-6) = 48$; -8 and -6 are a correct solution.
$(6)(8) = 48$; 6 and 8 are a correct solution.

PROBLEM 2: The width of a rectangular swimming pool is 10 feet less than the length of the swimming pool. The surface area of the pool is 600 square feet. What are the length and width of the sides of the pool?

First change the problem from Plain English to Math Talk. If the length of the pool is "l," then the width of the pool is "$l - 10$."

The area of any rectangle is length times width, so the area of the pool is $(l)(l - 10)$.

The area of the pool is 600 square feet.

So the problem can be written as $(l)(l - 10) = 600$.

To solve, multiply this equation.
$$l^2 - 10l = 600$$

Put the equation in standard form.
$$l^2 - 10l - 600 = 0$$
Factor this equation.
$$(l - 30)(l + 20) = 0$$

Solve for l.
If $l - 30 = 0$, then $l = 30$.
If $l + 20 = 0$, then $l = -20$.
The length of the pool cannot be a negative number, so the length must be 30 feet. If the length is 30 feet, then the width is $l - 10$, which is 20 feet.

Check your answer.
$(30)(20) = 600$ This is correct.
The length of the pool is 30 feet, and the width of the pool is 20 feet.

I'm sure this pool is ten feet longer than wide!

SUPER BRAIN TICKLERS

Solve for x.

1. $(x + 5)(x - 3) = 0$

2. $x^2 - 3x + 2 = 0$

3. $2x^2 - 3x - 2 = 0$

4. $x(x + 2) = -1$

5. $x^2 - 100 = 0$

6. $2x^2 = 50$

7. $3x^2 - 12x = 0$

8. $3x^2 - 4x = -1$

9. $3(x + 2) = x^2 - 2x$

10. $5(x + 1) = 2(x^2 + 1)$

(Answers are on page 250.)

BRAIN TICKLERS—
THE ANSWERS

Set # 43, page 208

1. $3x^2 + x = 0$

2. $2x^2 - 10x = 5$

3. $-2x^2 + 6x = 0$

4. $8x^2 - 12x = -3$

Set # 44, page 214

1. $x^2 + 7x + 10 = 0$

2. $x^2 - 2x - 3 = 0$

3. $6x^2 - 13x - 5 = 0$

4. $x^2 - 4 = 0$

Set # 45, page 216

1. $x^2 + 4x + 6 = 0$

2. $2x^2 - 3x + 3 = 0$

3. $5x^2 + 5x = 0$

4. $7x^2 - 7 = 0$

Set # 46, page 222

1. $x = 5; x = -5$

2. $x = 7; x = -7$

3. $x = 3; x = -3$

4. $x = 4; x = -4$

5. $x = \sqrt{15}; x = -\sqrt{15}$

6. $x = \sqrt{10}; x = -\sqrt{10}$

Set # 47, page 225

1. $x = 2; x = 0$

2. $x = -4; x = 0$

3. $x = 3; x = 0$

4. $x = -4; x = 0$

Set # 48, page 234

1. $x = -6, x = -4$

2. $x = 3, x = -4$

3. $x = \frac{5}{2}, x = 1$

4. $x = 3, x = -1$

Set # 49, page 244

1. $x = -3, x = -1$

2. $x = 2, x = -1$

3. $x = 1, x = 2$

4. $x = -\frac{1}{2}$

5. $x = 6, x = -6$

Super Brain Ticklers, page 247

1. $x = -5; x = 3$

2. $x = 1; x = 2$

3. $x = -\frac{1}{2}; x = 2$

4. $x = -1$

5. $x = 10; x = -10$

6. $x = 5; x = -5$

7. $x = 0; x = 4$

8. $x = \frac{1}{3}; x = 1$

9. $x = 6; x = -1$

10. $x = -\frac{1}{2}; x = 3$

CHAPTER EIGHT

Solving Systems of Equations

Some linear equations have more than one variable. In this chapter, you will learn how to solve equations with two variables. Here are examples of equations with two variables:

$$3x + 2y = 7$$
$$4x - y = -5$$
$$x - y = 0$$
$$2x - 2y = 6$$

It's impossible to solve a single equation with two variables and get a single solution. For example, in the equation $x - y = 0$, there are infinitely many possible answers. In fact, any time $x = y$, then $x - y = 0$. If $x = 5$ and $y = 5$, then $x - y = 0$. If $x = 100$ and $y = 100$, then $x - y = 0$.

But if you have two equations with two variables, such as $x + 3 = y$ and $x + y = 5$, there is one value of x and one value of y that will make both of these equations true. Pairs of equations with the same two variables are called *systems of linear equations*. Here are three examples of systems of linear equations:

$$x + y = 5 \ \text{ and } \ x - y = 5$$
$$2x - 3y = 7 \ \text{ and } \ 3x - 4y = 12$$
$$6x = 3y \ \text{ and } \ x + y = 3$$

In this chapter, you will learn how to solve systems of linear equations like these using three different techniques: addition or subtraction, substitution, and graphing.

SOLVING LINEAR EQUATIONS BY ADDITION OR SUBTRACTION

In order to solve a system of linear equations, first determine the relationship between the pair of equations. In a system of linear equations, there are three possible relationships between the pairs of equations.

RELATIONSHIP 1: In the pair of equations, the coefficient of one of the x terms is the opposite of the coefficient of the x term in the other equation. Or, in the pair of equations, the coefficient of one of the y terms is the opposite of the coefficient of the y term in the other equation. It sounds complicated but it's *painless*. Here is a pair of equations of this type.

$$x + 3y = 7$$
$$x - 3y = -2$$

Notice that in this pair of equations the coefficients in front of the y terms are 3 and -3; 3 and -3 are the opposite of each other. Here is another pair of equations of this type.

$$-6x + 2y = 0$$
$$6x + 4y = 3$$

In this pair of linear equations, the coefficients in front of the x terms are the opposite of each other; -6 is the opposite of 6.

When you find a pair of equations of this type it is easy to solve the system.

Just stick to the following steps.

Step 1: Add the two equations together.

Step 2: Solve the resulting equation.

Step 3: Substitute this answer back in one of the original equations to solve for the other variable.

Step 4: Check the answer.

Now watch as the following system of two equations is solved using addition. Notice that the coefficients in front of the y terms are the opposites of each other.

Solve $3x - 2y = 5$
$\qquad 3x + 2y = 13$

Step 1: Add the two equations together.
Watch what happens when you add these two equations together.

$$\begin{array}{r} 3x - 2y = 5 \\ 3x + 2y = 13 \\ \hline 6x = 18 \end{array}$$

The answer is one equation with one variable.

Step 2: Solve the resulting equation.

Solve $6x = 18$. Divide both sides of this equation by 6.

$$\frac{6x}{6} = \frac{18}{6}$$

Compute.

$$x = 3$$

Step 3: Substitute this answer back in one of the original equations to solve for the other variable.

Substitute $x = 3$ into the equation $3x - 2y = 5$.

The new equation is $3(3) - 2y = 5$.

Multiply $(3)(3) = 9$ and substitute 9 for $(3)(3)$.

The new equation is $9 - 2y = 5$.

Subtract 9 from both sides of the equation.

$$9 - 9 - 2y = 5 - 9$$

Compute.

$$-2y = -4$$

Now divide both sides of this equation by -2.

$$\frac{-2y}{-2} = \frac{-4}{-2}$$

Compute.

$$y = 2$$

Step 4: Check the answer.

The answer is $x = 3$ and $y = 2$.

Check this answer by substituting these values for both x and y in the original equations. The original two equations were $3x - 2y = 5$ and $3x + 2y = 13$.

Substitute $x = 3$ and $y = 2$ into $3x - 2y = 5$.

The resulting equation is $3(3) - 2(2) = 5$.

Multiply first.

$$(3)(3) = 9 \quad \text{and} \quad (2)(2) = 4.$$

Substitute these values into the equation.

$$9 - 4 = 5$$

This is a true sentence.

Now substitute $x = 3$ and $y = 2$ into the second equation, $3x + 2y = 13$.

The resulting equation is $3(3) + 2(2) = 13$.

Solve this equation. Multiply first: $(3)(3) = 9$ and $(2)(2) = 4$.

$$9 + 4 = 13$$

This is a true sentence.

So, the pair $x = 3$ and $y = 2$ makes $3x - 2y = 5$ *and* $3x + 2y = 13$ true.

Now watch as two more equations are solved using addition. Notice that the coefficients in front of the x variable are 1 and -1.

Solve $x + 4y = 17$
$-x - 2y = -9$

Step 1: Add the two equations together.
$$x + 4y = 17$$
$$\underline{-x - 2y = -9}$$
$$2y = 8$$

Step 2: Solve the resulting equation.
Solve $2y = 8$. Divide both sides by 2.
$$\frac{2y}{2} = \frac{8}{2}$$
$$y = 4$$

Step 3: Substitute this answer into one of the original equations to solve for the other variable.
Substitute $y = 4$ into the equation $x + 4y = 17$.
$$x + 4(4) = 17$$
Multiply.
$$(4)(4) = 16$$
Substitute 16 for $(4)(4)$.
$$x + 16 = 17$$
Subtract 16 from both sides of the equation.
$$x + 16 - 16 = 17 - 16$$
$$x = 1$$

Step 4: Check the answer.

The answer is $x = 1$ and $y = 4$.

Substitute this answer into the original equations. The original equations were $x + 4y = 17$ and $-x - 2y = -9$.

Substitute $x = 1$ and $y = 4$ into $x + 4y = 17$.

$$(1) + 4(4) = 17$$

Multiply.

$$(4)(4) = 16$$

Substitute 16 for (4)(4).

$$1 + 16 = 17$$

This is a true sentence.

Substitute $x = 1$ and $y = 4$ into the other original equation, $-x - 2y = -9$.

$$-(1) - 2(4) = -9$$

Multiply (2)(4).

$$(2)(4) = 8$$

Substitute.

$$-1 - 8 = -9$$

This is also a true sentence, so $x = 1$ and $y = 4$ make both equations true.

BRAIN TICKLERS
Set # 50

Solve the following systems of equations using addition.

1. $x - y = 4$

   $x + y = 8$

2. $3x + y = 0$

   $-3x + y = -6$

3. $2x + \frac{1}{4}y = -1$

   $x - \frac{1}{4}y = -2$

(Answers are on page 290.)

RELATIONSHIP 2: Sometimes two equations have the same coefficient in front of one of the variables. Following are two examples of equations with the same coefficients.

$$4x - y = 7$$
$$4x + 2y = 10$$

The coefficients in front of the x variable are both 4.

Same species, different equation.

$$3x - \frac{1}{2}y = 6$$

$$-2x - \frac{1}{2}y = -3$$

The coefficients in front of the y variable are both $-\frac{1}{2}$.
To solve equations with the same coefficient, follow the three *painless* steps in the following example.

Solve $\frac{1}{4}x + 3y = 6$

$\frac{1}{4}x + y = 4$

Step 1: Subtract one equation from the other.

$$\frac{1}{4}x + 3y = 6$$
$$- \left(\frac{1}{4}x + y = 4 \right)$$
$$\overline{}$$
$$2y = 2$$

Step 2: Solve the resulting equation.
Solve $2y = 2$.
Divide both sides of the equation by 2.

$$\frac{2y}{2} = \frac{2}{2}$$

Compute.

$$y = 1$$

Step 3: Substitute this answer into one of the original equations. Solve for the other variable.

Substitute $y = 1$ back into $\frac{1}{4}x + y = 4$.

$$\frac{1}{4}x + 1 = 4$$

Subtract 1 from both sides of the equation.

$$\frac{1}{4}x + 1 - 1 = 4 - 1$$

Simplify.

$$\frac{1}{4}x = 3$$

Multiply both sides of the equation by 4.

$$4\left(\frac{1}{4}x \right) = 4(3)$$

Simplify.

$$x = 12$$

Step 4: Check your answer.

Substitute $x = 12$ and $y = 1$ into the original equations.

Substitute $x = 12$ and $y = 1$ into $\frac{1}{4}x + 3y = 6$ and $\frac{1}{4}x + y = 4$.

Solve: $\frac{1}{4}(12) + 3(1) = 6$
$$3 + 3 \quad = 6$$
$$6 = 6$$

Solve: $\frac{1}{4}(12) + (6) \quad = 4$
$$3 + 1 \quad = 4$$
$$4 = 4$$

Here is an example of another pair of equations that have the same coefficient. Notice that both of these equations have a -2 in front of the y variable.

Solve $4x - 2y = 4$
$$-2x - 2y = 10$$

To solve these equations, follow the three-step *painless* solution.

Step 1: Subtract the second equation from the first equation.
$$4x - 2y = 4$$
$$-(-2x - 2y = 10)$$
$$\overline{6x \qquad = -6}$$

Step 2: Solve the resulting equation.

Solve $6x = -6$.

Divide both sides of the equation by 6.
$$\frac{6x}{6} = \frac{-6}{6}$$

Compute.
$$x = -1$$

Step 3: Substitute this answer into one of the original equations to solve for the other variable.

Substitute $x = -1$ into $4x - 2y = 4$
$$4(-1) - 2y = 4$$

Solve.
$$-4 - 2y = 4$$

Subtract.

$$-2y = 8$$

Divide both sides of the equation by -2.

$$\frac{-2y}{-2} = \frac{8}{-2}$$

Compute.

$$y = -4$$

Step 4: Check your answer.
Substitute $x = -1$ and $y = -4$ into $4x - 2y = 4$ and $-2x - 2y = 10$.

Solve: $4(-1) - 2(-4) = 4$
$-4 - (-8) = 4$
$4 = 4$

Solve: $-2(-1) - 2(-4) = 10$
$2 - (-8) = 10$
$10 = 10$

BRAIN TICKLERS
Set # 51

Solve these systems of equations using subtraction.

1. $3x - 2y = 7$

   $6x - 2y = 4$

2. $\frac{1}{2}x + 2y = 3$

   $\frac{1}{2}x - 5y = 10$

3. $3x + 6y = 9$

   $2x + 6y = 8$

4. $7x - \frac{2}{3}y = 12$

   $x - \frac{2}{3}y = 0$

(Answers are on page 290.)

RELATIONSHIP 3: Sometimes the coefficients of the two equations have no relationship to each other.

$$2x - 5y = 2$$
$$-5x + 3y = 4$$

Adding these two equations together would not help solve them.

Subtracting the second equation from the first will not help solve them either.

In order to solve equations of this type, follow these six *painless* steps.

Step 1: Multiply the first equation by the coefficient in front of the x in the second equation.

Step 2: Multiply the second equation by the coefficient in front of the x of the first equation.

Step 3: Add or subtract these two new equations.

Step 4: Solve the resulting equation.

Step 5: Substitute this answer back into the original equation to solve for the other variable.

Step 6: Check your answer.

Now watch as these two equations are solved.

Solve $2x - 5y = 2$
$-5x + 3y = 4$

Step 1: Multiply the first equation by the coefficient in front of the x in the second equation.
Negative five is the coefficient in front of the second equation.
Multiply the first equation by -5.
$$(-5)(2x - 5y = 2)$$
$$-5(2x) - 5(-5y) = -5(2)$$
$$-10x + 25y = -10$$

Step 2: Multiply the second equation by the coefficient in front of the x of the first equation.
Two is the coefficient in front of the first equation.
Multiply the second equation by two.
$$(2)(-5x + 3y = 4)$$
$$2(-5x) + 2(3y) = 2(4)$$
$$-10x + 6y = 8$$

Step 3: Add or subtract these two new equations.
Subtract these two new equations.
$$\begin{array}{r} -10x + 25y = -10 \\ -10x + 6y = 8 \\ \hline 19y = -18 \end{array}$$

Step 4: Solve the resulting equation.
$$19y = -18$$
Divide both sides of this equation by 19.
$$\frac{19y}{19} = \frac{-18}{19}$$
$$y = \frac{-18}{19}$$

Step 5: Substitute this answer back into the original equation to solve for the other variable.

$$2x - 5\left(\frac{-18}{19}\right) = 2$$

Multiply.

$$2x + \frac{90}{19} = 2$$

Subtract.

$$2x = 2 - \frac{90}{19}$$

$$2x = \frac{38}{19} - \frac{90}{19}$$

$$2x = \frac{-52}{19}$$

Divide both sides by 2.

$$\frac{2x}{2} = \left(\frac{-52}{19}\right) \div 2$$

$$x = \frac{-26}{19}$$

Step 6: Check your answer.

$$x = \frac{-26}{19} \text{ and } y = \frac{-18}{19}$$

Check these answers for practice by substituting them back into both of the original equations.

Here is another solution of a system of two equations.

Solve $3x - 2y = 9$
$-x + 3y = 4$

Adding these two equations together would not help solve them. Subtracting these two equations would not help solve them either. Use the six-step process for solving equations with no relationship.

Step 1: Multiply the first equation by the coefficient in front of the x in the second equation.
Negative one is the coefficient in front of the x in the second equation. Multiply the first equation by negative one.
$(-1)(3x - 2y = 9)$ is $-3x + 2y = -9$.

Step 2: Multiply the second equation by the coefficient in front of the x of the first equation.
Three is the coefficient in front of the first equation.
Multiply the second equation by three.
$$(3)(-x + 3y = 4) \text{ is } -3x + 9y = 12.$$

Step 3: Add or subtract these two new equations.
Subtract these two new equations.

$$
\begin{array}{r}
-3x + 2y = -9 \\
-3x + 9y = 12 \\
\hline
-7y = -21
\end{array}
$$

Step 4: Solve the resulting equation.
$$\text{If } -7y = -21, y = 3.$$

Step 5: Substitute the value of y into the original equation to solve for x.
Substitute 3 for y in the equation $3x - 2y = 9$.
$$3x - 2(3) = 9$$
Multiply.
$$3x - 6 = 9$$
Solve.
$$3x = 15$$
$$x = 5$$

Step 6: Check. Substitute the values of x and y into one of the original equations. If the result is a true sentence, the equation is correct.
Check this solution on your own.

BRAIN TICKLERS
Set # 52

Multiply the first equation by the coefficient of the x term of the second equation. Multiply the second equation by the coefficient of the x term of the first equation. Next, add or subtract to solve these equations.

1. $3x + 2y = 12$

   $x - y = 10$

2. $2x - y = 3$

   $-4x + y = 6$

3. $-5x + y = 8$

   $-2x + 2y = 4$

4. $\frac{1}{2}x - 2y = 6$

   $4x + 2y = 12$

(Answers are on page 290.)

SOLVING LINEAR EQUATIONS BY SUBSTITUTION

Five painless steps to solution point!

Substitution is another way to solve a system of linear equations. Here is how you solve two equations with two variables using substitution.

Follow these *painless* steps.

Step 1: Solve one of the equations for x. The answer will be in terms of y.

Step 2: Substitute this value for x in the other equation. Now there is one equation with one variable.

Step 3: Solve the equation for y.

Step 4: Substitute the value of y into one of the original equations to find the value of x.

Step 5: Check. Substitute the values of both x and y into both of the original equations. If each result is a true sentence, the solution is correct.

Watch as this system of two linear equations is solved using substitution.

Solve $x - y = 3$
$\qquad$ $2x + y = 12$

Step 1: Solve one of the equations for x. The answer will be in terms of y.
Solve $x - y = 3$ for x.
Add y to both sides of the equation.
$$x - y + y = 3 + y$$
Simplify.
$$x = 3 + y$$

Step 2: Substitute this value for x in the other equation. Now there is one equation with one variable.
Substitute $(3 + y)$ into $2x + y = 12$ wherever there is an x.
$$2(3 + y) + y = 12$$

Step 3: Solve the equation for y.

Use the order of operations to simplify the equation.

$$6 + 2y + y = 12$$

Simplify.

$$6 + 3y = 12$$

Subtract 6 from both sides of the equation.

$$6 + 3y - 6 = 12 - 6$$

Simplify.

$$3y = 6$$

Divide both sides of this equation by 3.

$$\frac{3y}{3} = \frac{6}{3}$$

Simplify.

$$y = 2$$

Step 4: Substitute the value of y into one of the original equations to find the value of x.

Substitute $y = 2$ into the equation $x - y = 3$

$$x - 2 = 3$$

Solve this equation for x. Add 2 to both sides of the equation.

$$x - 2 + 2 = 3 + 2$$

Simplify.

$$x = 5$$

Step 5: Check. Substitute the values of both x and y into the original equations. If each result is a true sentence, the solution is correct.

Substitute $x = 5$ and $y = 2$ into $2x + y = 12$ to check the answer.

$$2(5) + 2 = 12$$

Compute the value of this expression.

$$10 + 2 = 12$$
$$12 = 12$$

This is a true sentence. You can also check that the values $x = 5$ and $y = 2$ make the equation $x - y = 3$ true. The solution $x = 5$ and $y = 2$ is correct.

Watch as two more linear equations are solved using substitution.

Solve $x + 3y = 6$
$\quad\quad\quad x - 3y = 0$

Step 1: Solve one of the equations for x. The answer will be in terms of y.
Solve $x - 3y = 0$ for x.
Add $3y$ to both sides of the equation.
$$x - 3y + 3y = 0 + 3y$$
Simplify.
$$x = 3y$$

Step 2: Substitute this value for x in the other equation.
Substitute $x = 3y$ into the equation $x + 3y = 6$.
$$3y + 3y = 6$$
Now there is one equation with one variable.

Step 3: Solve the equation $3y + 3y = 6$ for y.
Simplify.
$$3y + 3y = 6$$
Combine like terms.
$$6y = 6$$
Divide both sides of the equation by 6.
$$\frac{6y}{6} = \frac{6}{6}$$
Simplify.
$$y = 1$$

Step 4: Substitute the value of y into one of the original equations to find the value of x.
Substitute $y = 1$ into the equation $x + 3y = 6$.
$$x + 3(1) = 6$$
Now solve for x.
$$x + 3 = 6$$
Subtract 3 from both sides of the equation.
$$x + 3 - 3 = 6 - 3$$
Simplify.
$$x = 3$$

Step 5: Check. Substitute the values of both x and y into both of the original equations. If each result is a true sentence, the solution is correct.

$$x = 3 \text{ and } y = 1$$

Substitute these numbers in the equation $x - 3y = 0$.

$$3 - 3(1) = 0$$
$$3 - 3 \quad = 0$$

This is a true sentence. You can also check that the values $x = 3$ and $y = 1$ make the equation $x + 3y = 6$ true.

The solution $x = 3$ and $y = 1$ is correct.

BRAIN TICKLERS
Set # 53

Solve the following systems of equations using substitution.

1. $x + y = 7$

   $x - y = 1$

2. $2x + 5y = 7$

   $x + y = 2$

3. $2x + y = 0$

   $x + y = -2$

4. $2x + y = 3$

   $4x + 3y = 8$

(Answers are on page 290.)

SOLVING LINEAR EQUATIONS
BY GRAPHING

Graphing is the third way to solve a pair of linear equations. Before you can learn how to solve a linear equation by graphing, you need to learn how to plot points and graph a line.

Imagine two number lines that intersect each other. One of these lines is horizontal and the other line is vertical. The horizontal line is called the *x*-axis. The vertical line is called the *y*-axis. Where the lines intersect is called the *origin*. The origin is the point (0, 0). Each of these two number lines has numbers on it.

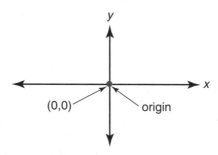

Look at the x-axis. It is just like a number line. The numbers to the right of the origin are positive. The numbers to the left of the origin are negative.

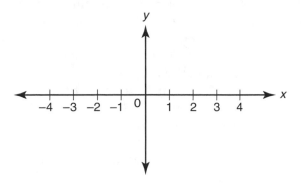

Look at the y-axis. It is like a number line that is standing straight up. The numbers on the top half of the number line are positive. The numbers on the bottom half of the number line are negative.

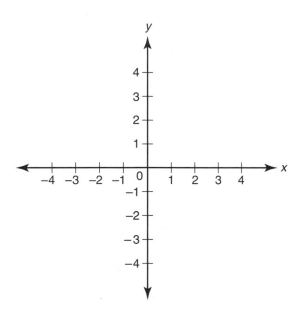

Graphing points

You can graph points on this coordinate axis system.
Points to be graphed are written as two numbers, such as (3, 2).
The first number is the x value. The second number is the y
value. The first number tells you how far to the left or the right
of the origin the point is. The second number tells you how far
above or below the origin the point is.

To graph a point, follow these four *painless* steps.

Step 1: Put your pencil at the origin.

Step 2: Start with the x term. It is the first term in the
parentheses.

Move your pencil x spaces to the left if the x term is
negative.
Move your pencil x spaces to the right if the x term is
positive.
Keep your pencil at this point.

Step 3: Look at the y term. It is the second term in the
parentheses.

Move your pencil y spaces down if the y term is negative.
Move your pencil y spaces up if the y term is positive.

Step 4: Mark this point.

Graph (4, 2).
To graph the point (4, 2), put your pencil at the origin.
Move your pencil four spaces to the right.
Move your pencil two spaces up.
Mark this point. This is the point (4, 2).

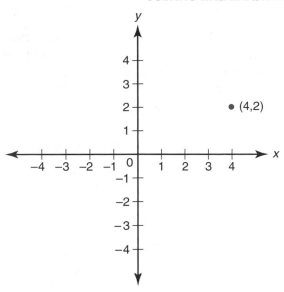

Graph (−3, −1).
To graph the point (−3, −1), put your pencil at the origin.
Move your pencil three spaces to the left.
Move your pencil one space down.
Mark this point. This is the point (−3, −1).

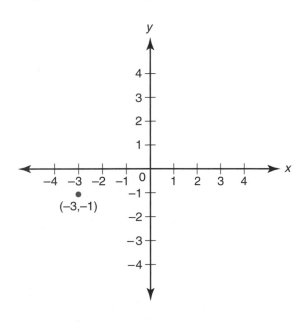

Graph (−2, 0).

Negative two is the x value and zero is the y value.
To graph the point $(-2, 0)$, put your pencil at the origin.
Move your pencil two spaces to the left.
Do not move your pencil up or down, since the y value is 0.
The point $(-2, 0)$ lies exactly on the x-axis.

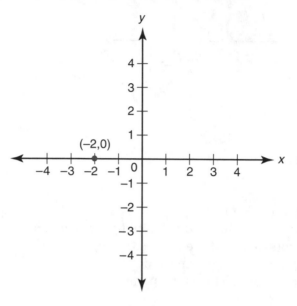

Graph (0, 3).

Zero is the x value and three is the y value. Start at the origin. Since the x value is zero, do not move to the left or the right. Since three is the y value, move up three spaces. The point $(0, 3)$ lies directly on the y-axis.

BRAIN TICKLERS
Set # 54

Graph the following points.

1. $(1, 4)$ 5. $(0, 3)$

2. $(3, -1)$ 6. $(4, 0)$

3. $(-2, 6)$ 7. $(0, 0)$

4. $(-4, -2)$

(Answers are on page 291.)

Graphing lines

Now that you know how to graph individual points, you can graph a linear equation. The graph of a linear equation is a straight line. To graph a linear equation, find three points that make an equation true. Connect these points into a straight line.

To graph a linear equation, follow these four *painless* steps.

Step 1: Solve the equation for y.

Step 2: Find three points that make the equation true.

Step 3: Graph the three points.

Step 4: Connect the three points into a straight line. Make sure to extend the line with arrows to show that it goes on forever.

Graph $x - y + 1 = 0$.
To graph the linear equation $x - y + 1 = 0$, follow the four *painless* steps.

Step 1: Solve the equation for y.
$$x - y + 1 = 0$$
Add y to both sides of the equation.
$$x - y + 1 + y = 0 + y$$
Simplify by combining like terms.
$$x + 1 = y$$

Step 2: Find three points that make the equation true. Pick a number for x and figure out the corresponding y value.

If $x = 0$, $y = 1$. The point $(0, 1)$ makes the equation $y = x + 1$ true.
If $x = 1$, $y = 2$. The point $(1, 2)$ also makes this equation true.
If $x = 2$, $y = 3$. The point $(2, 3)$ also makes this equation true.

Step 3: Graph the three points.
Graph (0, 1), (1, 2), and (2, 3).

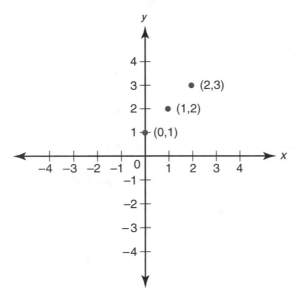

Step 4: Connect and extend the three points to make a straight line.
This is the graph of the equation $x - y + 1 = 0$.

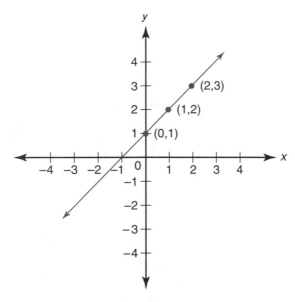

Graph $2x + y = 4$.

To graph the linear equation $2x + y = 4$, follow the four *painless* steps.

Step 1: Solve the equation for y.

Subtract $2x$ from both sides of the equation.
$$2x - 2x + y = 4 - 2x$$

Simplify.
$$y = 4 - 2x$$

Step 2: Find three points that make the equation true.

If $x = 0$, $y = 4$. The point $(0, 4)$ makes the equation $y = 4 - 2x$ true.
If $x = 1$, $y = 2$. The point $(1, 2)$ makes the equation $y = 4 - 2x$ true.
If $x = 2$, $y = 0$. The point $(2, 0)$ makes the equation $y = 4 - 2x$ true.

Step 3: Graph the three points.

Graph $(0, 4)$, $(1, 2)$, and $(2, 0)$.

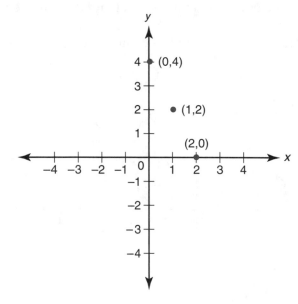

Step 4: Connect and extend the three points to make a straight line.

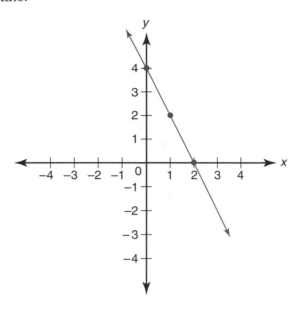

You have graphed the equation $2x + y = 4$.

Graph $x - y = 0$.
To graph the linear equation $x - y = 0$, follow the four *painless* steps.

Step 1: Solve the equation $x - y = 0$ for y.
Add y to both sides of the equation $x - y = 0$.
$$x - y + y = 0 + y$$
Simplify by combining like terms.
$$x = y$$

Step 2: Find three points that make the equation true.

If $x = 0$, $y = 0$. The point $(0, 0)$ makes the equation
$x - y = 0$ true.
If $x = 1$, $y = 1$. The point $(1, 1)$ makes this equation true.
If $x = 4$, $y = 4$. The point $(4, 4)$ makes this equation true.

Step 3: Graph the three points.

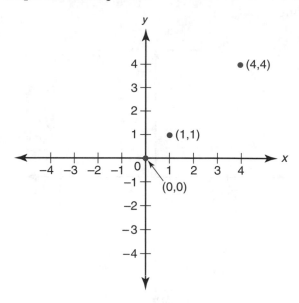

Step 4: Connect and extend the three points to make a straight line.

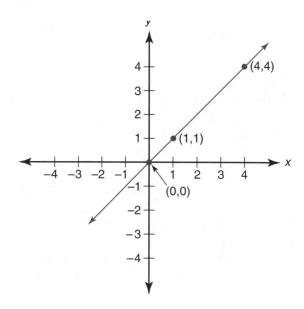

You have graphed the equation $x - y = 0$.

Solving a system of linear equations by graphing

To solve a system of linear equations by graphing, follow these four steps.

Step 1: Graph the first equation.

Step 2: Graph the second equation on the same set of axes.

Step 3: Find the solution, which is the point where the two lines intersect.

Step 4: Check the answer. Substitute the intersection point into each of the two original equations. If each equation is true, the answer is correct.

Watch as graphing is used to solve a system of two linear equations.

Solve $2x + y = 1$ and $x - y = -1$.

Step 1: Graph the first equation, $2x + y = 1$.
Solve the equation for y.
Subtract $2x$ from both sides of the equation.
$$2x - 2x + y = 1 - 2x$$
Simplify.
$$y = 1 - 2x$$

Now find three points by substituting 0, 1, and 2 for x.

If $x = 0$, $y = 1$.
If $x = 1$, $y = -1$.
If $x = 2$, $y = -3$.

Now graph these points. Connect and extend them.

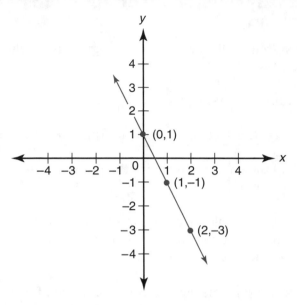

Step 2: Graph the second equation, $y - 1 = x$, on the same
axes. Solve the equation $y - 1 = x$ for y.
Add 1 to both sides of the equation.
$$y - 1 + 1 = x + 1$$
Simplify.
$$y = x + 1$$
Now find three points that make the equation $y = x + 1$ true.

If $x = 0$, $y = 1$.
If $x = 1$, $y = 2$.
If $x = 2$, $y = 3$.

Graph these points and connect them on the same
graph that the first equation was graphed on.

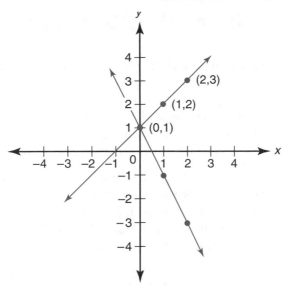

Step 3: The solution is the point where the two lines intersect. These lines intersect at the point (0, 1).

Step 4: Check the answer. Substitute the intersection point into each of the two original equations. If both sentences are true, then the answer is correct.
Substitute the point (0, 1) into the equation $2x + y = 1$.
Substitute 0 for x and 1 for y.

$$2(0) + 1 = 1$$

Compute.

$$1 = 1$$

Substitute the point (0, 1) into the equation $x - y = -1$.
Substitute 0 for x and 1 for y.

$$0 - 1 = -1$$

Compute.

$$-1 = -1$$

Both sentences are true.
The solution is correct.

WORD PROBLEMS

Watch how systems of equations can be used to solve word problems.

PROBLEM 1: Together Keisha and Martha earned $8.
Keisha earned $2 more than Martha.
How much money did Keisha and Martha each earn?

To solve this word problem, change it into two equations.
Pick two letters to represent Keisha and Martha.
Let K represent Keisha and M represent Martha.
Change each sentence into an equation.
Together, Keisha and Martha earned $8.
$K + M = 8$
Keisha earned $2 more than Martha.
$K - M = 2$

Solve these two equations using addition.

$$\begin{array}{rcl} K + M &=& 8 \\ K - M &=& 2 \\ \hline 2K &=& 10 \end{array}$$

Divide both sides of the equation $2K = 10$ by 2.
$\frac{2K}{2} = \frac{10}{2}$
Compute.
$K = 5$

Substitute 5 back into the original equation, $K + M = 8$ to solve for M.

$5 + M = 8$

Subtract 5 from both sides of the equation.

$5 + M - 5 = 8 - 5$

Simplify.

$M = 3$

The answers are $K = 5$ and $M = 3$.

To check these answers, substitute these numbers in the other equation.

Substitute $K = 5$ and $M = 3$ into the equation $K - M = 2$.

$5 - 3 = 2$

This is a true sentence.

Keisha earned \$5, and Martha earned \$3.

PROBLEM 2: Jorge is twice as old as Sean.
Together, Jorge and Sean have lived 18 years.
How old are Jorge and Sean?

To solve this word problem, change it into two equations.
Pick two letters to represent Jorge and Sean.
Let J represent Jorge and S represent Sean.
Change each sentence into an equation.

Jorge is twice as old as Sean.
$J = 2S$
Together, Jorge and Sean have lived 18 years.
$J + S = 18$

Watch as these two equations are solved by substitution.
$J = 2S$
$J + S = 18$
Substitute $2S$ for J in the equation $J + S = 18$.
$2S + S = 18$
Simplify.
$3S = 18$
Divide both sides by 3.
$$\frac{3S}{3} = \frac{18}{3}$$
Simplify.
$S = 6$

Substitute 6 for S into the original equation $J = 2S$ to solve for J.
$J = 2(6)$
Multiply.
$J = 12$

Substitute $S = 6$ and $J = 12$ into the other original equation, $S + J = 18$.
$6 + 12 = 18$
Simplify.
$18 = 18$
The answers are correct.
Sean is 6 years old and Jorge is 12 years old.

SUPER BRAIN TICKLERS

Solve the following systems of equations using addition.

1. $x + y = 12$

   $2x - y = 0$

2. $-2x + 5y = 1$

   $x - 2y = 4$

Solve the following systems of equations using substitution.

3. $3x - y = 4$

   $-2x + y = 1$

4. $x + y = 7$

   $3x - 2y = 4$

(Answers are on page 291.)

BRAIN TICKLERS—
THE ANSWERS

Set # 50, page 258

1. $x = 6$; $y = 2$

2. $x = 1$; $y = -3$

3. $x = -1$; $y = 4$

Set # 51, page 262

1. $x = -1$; $y = -5$

2. $x = 10$; $y = -1$

3. $x = 1$; $y = 1$

4. $x = 2$; $y = 3$

Set # 52, page 267

1. $x = 6\frac{2}{5}$, $y = -\frac{18}{5}$

2. $x = -\frac{9}{2}$, $y = -12$

3. $x = -\frac{3}{2}$, $y = \frac{1}{2}$

4. $x = 4$, $y = -2$

Set # 53, page 272

1. $y = 3$; $x = 4$

2. $x = 1$; $y = 1$

3. $x = 2$; $y = -4$

4. $x = \frac{1}{2}$; $y = 2$

Set # 54, page 277

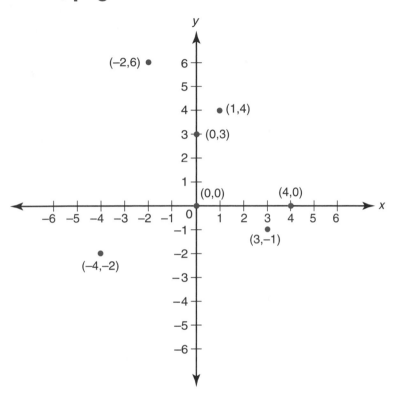

Super Brain Ticklers, page 289

1. $x = 4$; $y = 8$

2. $x = 22$; $y = 9$

3. $x = 5$; $y = 11$

4. $x = \frac{18}{5}$; $y = \frac{17}{5}$

INDEX

Really. This isn't going to hurt at all . . .

BARRON'S

Barron's *Painless* titles are perfect ways to show kids in middle school that learning really doesn't hurt. They'll even discover that grammar, algebra, and other subjects that many of them consider boring can become fascinating— and yes, even fun! The trick is in the presentation: clear instruction, taking details one step at a time, adding a light and humorous touch, and sprinkling in some brain-tickler puzzles that are both challenging and entertaining to solve.

Each book: Paperback, approx. 224 pp., $8.95–$10.95, Canada $11.95–$15.95

Painless Algebra
Lynette Long, Ph.D.
ISBN 0-7641-0676-7

Painless American Government
Jeffrey Strausser
ISBN 0-7641-2601-6

Painless American History
Curt Lader
ISBN 0-7641-0620-1

Painless Fractions
Alyece Cummings
ISBN 0-7641-0445-4

Painless Geometry
Lynette Long, Ph.D.
ISBN 0-7641-1773-4

Painless Grammar
Rebecca S. Elliott, Ph.D.
ISBN 0-8120-9781-5

Painless Math Word Problems
Marcie Abramson, B.S., Ed.M.
ISBN 0-7641-1533-2

Painless Poetry
Mary Elizabeth
ISBN 0-7641-1614-2

Painless Reading Comprehension
Darolyn E. Jones
ISBN 0-7641-2766-7

Painless Research Projects
*Rebecca S. Elliott, Ph.D.,
and James Elliott, M.A.*
ISBN 0-7641-0297-4

Painless Science Projects
Faith Hickman Brynie, Ph.D.
ISBN 0-7641-0595-7

Painless Speaking
Mary Elizabeth
ISBN 0-7641-2147-2

Painless Spelling
Mary Elizabeth
ISBN 0-7641-0567-1

Painless Writing
Jeffrey Strausser
ISBN 0-7641-1810-2

Barron's Educational Series, Inc.
250 Wireless Boulevard, Hauppauge, NY 11788
In Canada: Georgetown Book Warehouse
34 Armstrong Avenue, Georgetown, Ont. L7G 4R9
Visit our website @ www.barronseduc.com

Prices subject to change without notice. Books may be purchased at your local bookstore, or by mail from Barron's. Enclose check or money order for total amount plus sales tax where applicable and 18% for shipping and handling ($5.95 minimum). NY State and California residents add sales tax.

(#79) 6/04